SpringerBriefs in Applied Sciences and Technology

PoliMI SpringerBriefs

Springer, in cooperation with Politecnico di Milano, publishes the PoliMI Springer-Briefs, concise summaries of cutting-edge research and practical applications across a wide spectrum of fields. Featuring compact volumes of 50 to 125 (150 as a maximum) pages, the series covers a range of contents from professional to academic in the following research areas carried out at Politecnico:

- Aerospace Engineering
- Bioengineering
- Electrical Engineering
- Energy and Nuclear Science and Technology
- Environmental and Infrastructure Engineering
- Industrial Chemistry and Chemical Engineering
- Information Technology
- Management, Economics and Industrial Engineering
- Materials Engineering
- Mathematical Models and Methods in Engineering
- Mechanical Engineering
- Structural Seismic and Geotechnical Engineering
- Built Environment and Construction Engineering
- Physics
- Design and Technologies
- Urban Planning, Design, and Policy

Daniele Fanzini · Gianpiero Venturini

Reactivation of the Built Environment

From Theory to Practice

POLITECNICO
MILANO 1863

Daniele Fanzini (ID)
Rivergaro, Italy

Gianpiero Venturini
Madrid, Spain

ISSN 2191-530X   ISSN 2191-5318 (electronic)
SpringerBriefs in Applied Sciences and Technology
ISSN 2282-2577   ISSN 2282-2585 (electronic)
PoliMI SpringerBriefs
ISBN 978-3-031-16068-4   ISBN 978-3-031-16069-1 (eBook)
https://doi.org/10.1007/978-3-031-16069-1

This Springer imprint is published by the registered company Springer Nature Switzerland AG
The registered company address is: Gewerbestrasse 11, 6330 Cham, Switzerland

Preface

The economic–financial crisis in 2008 and the pandemic in 2020, have placed the issue of rehabilitation of the built environment back at the center of the urban and architectural debate, with particular emphasis on social, environmental and feasibility issues. In this sense, the crisis represents an important watershed, as it not only brought about new economic, political, social and technological conditions, such as restoring the focus on the built environment, but also encouraged the spread of alternative modes of intervention to the existing ones. These include forms of urban regeneration via social innovation designed to respond to the actual problems of civil society, which traditional public and private actors can no longer cope with. Urban reactivation shares this approach, focusing on the many unused urban spaces which, due to their characteristics, don't match the interests of the private/public operators.

This book deals with the theme of urban reactivation as a particular form of regeneration intervention, which in addition to the physical-spatial dimension of the places also—and above all—considers the social and relational dynamics that the intervention is able to activate. In this sense, the concept of activation (or reactivation) emphasizes the act of putting something into or back into operation, whether it concerns the material components of a container (a building or a place) or the immaterial components of a content (a need, a function), starting from the exploitation of opportunities that the architectural project contributes to revealing, developing and accompanying. At the center of the topic are public and/or private spaces and buildings that have lost their original purpose—or never had one—and are in need of actions to define or redefine their uses in relation to the needs of today's world. These spaces represent important territorial resources from which to activate the local community and trigger transformations capable of producing socio-spatial effects.

The book explains the way in which urban reactivation interventions are carried out and describes the possible contribution of the project as a tool to promote the success of such interventions by foreseeing the social, procedural and morphological frameworks of possible strategies. The analysis of case studies, as well as carrying out field activities such as interviews and reactivation project initiatives conducted in the real world, introduce readers to the characteristics of the reactivation intervention and the useful tools for managing its project implications.

In summary, the first part of the book investigates the reasons why, in recent years, there has been an increased focus on urban regeneration issues, also addressing the taxonomic aspects related to the definition of the reactivation intervention and its characteristics of action. The methodology, in line with the principles of the place-based approach and the principles of integrated urban development promoted by European policies, adopts the semantic shift introduced by urban regeneration processes via social innovation.

In the second part, the book presents the methodology of urban reactivation as a new approach to urban regeneration. In this new approach, the scale of the project is not a decisive factor, as the approach favors the ways in which the design choices of interventions mature and the social ramifications, they are able to produce. In this sense, the book examines a different form of scalarity, which is linked to effects rather than the actual size of interventions. This is a multiple scalarity that operates on the vertical relationships between the local and global dimensions, but also and above all, on the horizontal connections between the urban spaces.

The third and last part of the book focuses on the applicational aspects of the method, deepening three enabling factors:

(a) the key role of the planner, seen as an activator and mediator of the requests of the subjects variously involved in the reactivation process. A genuinely enabling figure capable of acting as an agent of social, cultural and political mediation who acts not only as a creator of form, but also as an enabler of processes and narratives;

(b) the open and flexible nature of urban regeneration processes in fostering the emergence of new economies at the microscale of neighborhoods, which in turn can trigger forms of participation, encouraging inclusiveness and a sense of belonging on the part of the communities involved;

(c) the need for technological tools, including digital ones, capable of facilitating the involvement of a wide range of users, from the recognition of a problem to its practical treatment in design and construction terms.

The cognitive investigation of these three aspects has led to the definition of the methodological and instrumental apparatus of the urban regeneration intervention, which was subsequently tested through real-world experiences. A collection of selected interviews, case studies and field trials provide references for understanding the application of the method examined.

Rivergaro, Italy   Daniele Fanzini
Madrid, Spain   Gianpiero Venturini

Acknowledgements

This book summarizes some of the results of extensive research carried out by the authors with the aim of defining the theory and practice of urban reactivation of the built environment and the competences of the urban reactivator as a professional figure of reference. This activity led the authors to deal with numerous experts and professionals, who with great generosity and patience made their experience available, contributing significantly to the achievement of this first result. Their names are listed in the third section of the book dedicated to interviews and case study analysis. Our heartfelt thanks go to them.

Contents

Chapter 1
Intervention on the Built Environment: Approaches and Strategies for the City

Abstract The chapter presents the theoretical aspects that led to the urban reactivation definition and conceptualization. The analysis contains premises for the definition of the project methodology described in Chap. 2 as well as for the identification of the theoretical and practical issues that are still unresolved. In summary, the chapter investigates the reasons why, in recent years, there has been an increased focus on urban regeneration issues and the innovative actions and processes implemented. It then addresses the nomenclature aspects related to urban transformation and regeneration interventions and the characteristics of the action. Finally, the chapter investigates the principles of urban reactivation, those related to social innovation, meaning the use of forms of sharing and cooperation to solve problems and satisfy the common needs of society. This first chapter frames the theoretical aspects of research that is halfway between the inductive and deductive method combining theoretical research, the analysis of case studies, as well as carrying out field activities such as interviews and reactivation project initiatives conducted in the real world described in Chap. 3. All these elements outline, through a process of progressive refinement of knowledge, the characteristics of urban reactivation and the useful tools for managing its implications in the project.

1.1 From the Crises: Reasons for a New Approach to the Built Environment

The economic and financial crisis of 2008 has had the merit, if one can say so, of exposing the great weakness of a development model based on continuous economic growth. This is a model which in Italy, in the period from 1994 to 2006 alone, resulted in the production of around a billion cubic meters of new residential buildings, saturating the market and causing a significant drop in demand. It is estimated that in the five-year period 2006–2011 alone, the number of house purchases and sales fell by 31.3% in terms of quantity, and 21.5% in terms of price (CRESME 2012). Although there has been a partial recovery in real estate values over the years, a substantial proportion of pre-crisis production has remained unsold worsening the phenomenon of disposal, understood as the process of partial or total decommissioning of areas, agglomerations, and buildings intended for productive activities (Dansero 1993).

Unsold and abandoned properties are the cause of the slow but relentless increase in the presence of public and private properties (residential buildings, areas, capital goods, infrastructures) in a state of abandonment in western cities, and the tangible and evident sign of the considerable changes that the construction sector is still going through. For this reason, in the early 1990s, also as a consequence of European projects such as 'Urban Pilot' and 'Urban Initiative', the term 'urban regeneration' began to circulate like an ambiguous locution (Jones and Evans 2013) used to indicate actions on the built environment having the aim of accompanying the physical dimension of property recovery with the solution of social problems, enhancing, in a context of integrated development, the available local resources (Saccomani 2016).

Roberts et al. (2017) define 'urban regeneration' as a mix of vision and action on a place-based basis with the aim of solving urban problems while producing lasting improvements in social, economic and environmental conditions in a given place at a given time. In this sense, the place represents the origin and destination of the strategies and actions for improvement: origin as the source of the opportunities, destination as the route for the territorialization of rehabilitation policies and strategies. The term 'territorialization of approaches' indicates the possibility of increasing the added value of policies by involving the system of local operators and promoting concrete and realistic projects conceived as an overall urban development strategy (Pilot Urban Projects/Progetti Urbani Pilota, established by the ERDF in 1989).

The variety of approaches adopted to manage the complex physical, social, economic and political implications (Roberts et al. 2017) of built environment rehabilitation will lead to a rather broad and articulated picture of declination of the urban regeneration term, all characterized by the paradigm shift imposed by the fact that the public sector is no longer able to address social, urban and environmental issues as intensively as it used to, and is therefore in need of economic and operational support.

The economic crisis has amplified the inability of public actors to reduce episodes of neglect and abandonment, both because it has reduced the economic resources available, and because, as Perriccioli (2017) notes, it has marked the decline of the ideals of equality and freedom that govern the delicate balance between the different parts of the city. Urban regeneration can have a top-down or bottom-up trigger, but in any case, needs planning resources of a different nature, and acts through integrated formulas that combine the physicality of places with the immateriality of social, economic, environmental and cultural aspects (Galdini 2008).

Ermacora and Bullivant (2016) highlight a problematic gap between people's aspirations for self-governance and the ways in which land governance operates. The boundary between soft and hard, short and long planning is more blurred than it used to be, and the proposed solutions need to be targeted at the specific problem. This leads to the overturning of traditional practices in which politics and strategy prevail over action, and with it the need for new professional codes. In this sense, Venturini and Venegoni (2016) adopts the term of urban reactivation to indicate "new models that currently lack connections to theoretical practice, and, as a response, a number of initiatives employing more experimental strategies and methodologies have resulted in, quite interestingly, a sort of systematization of the informal". If one of the fundamental aspects of urban regeneration is the political will that results in a long-term collaborative vision (Roberts et al. 2017; Saccomani 2016), in urban reactivation the political will originates from the single asset and from the opportunities that this asset is able to offer to the network of economic and social operators in the territory.

Under the term 'civic entrepreneurs' Henton et al. (1997) define the set of actors capable of creating the conditions to achieve a common goal, which in the case of urban regeneration most often originates from the chance to reuse an abandoned or disused property. This is like the 'space-feeling-action' concept used by Fanzini and Rotaru (2018a) to signify the commonality of intent between public and private that is activated through mixed approaches straddling top-down and bottom-up formulas of urban intervention. In the urban reactivation approach, the need for new strategies of intervention on the city starts from the living experience of the inhabitants and improves thanks to the collaborative sense of the relationship with politics and with the traditional professional players, such as planners, designers. This leads to the need for flexible tools able to adapt to the variability of the relationship between strategy and action, experimentation and decision, and to a new skill able to mobilize local resources, including those of a relational nature, to trigger new hypotheses of growth and development (Capasso and Georgieff 2015; La Cecla 2015; Ostanel 2017).

The motivations for the project thus conceived derive primarily from the need to respond to the new social demands mentioned above, but also those dictated by higher order collective interests, and which have to do with the quality of the collective living environment. It is precisely the inhabitants of the contemporary city who have shown greater awareness of environmental, social and cultural issues, and it is precisely these issues that call for new tools of dialogue and mediation in urban transformation interventions, including reactivation (Manzini 2018).

1.2 The Built Environment Transition

After the expansive phase of the 1980s, cities began to implode and show clear signs of decay and abandonment due to the transition from a manufacturing economy to a post-industrial economy based on services and information (Galdini 2008) and the consequent change in strategic urban functions. Smets (1990) effectively describes this transition through what he refers to as a 'taxonomy of neglect', i.e., a description of the processes of growth and decay that have shaped cities, and the strategies involved in controlling these changes.

As Valente (2017) notes, Smets' contribution highlights two fundamental points for interpreting the phenomenon of urban divestment: the ways in which functions or uses of space are removed and the parallel definition of the urban void as a waiting space full of potential. What emerges is a picture that is still valid today and provides an understanding of how entire territories have changed and the practices that have produced the highest quality environmental and social interventions. These practices are generally oriented towards growth, not in purely additive terms (expanding the city by occupying agricultural land), but in terms of re-functioning and re-signifying existing spaces. This applies both to large urban transformation interventions managed through top-down approaches and dedicated public policies—the same policies that in northern Europe have regenerated large portions of early industrialized territory—and through more subtle bottom-up practices of urban recycling, such as temporary reuse, tactical urbanism, functional reuse, urban retrofitting, placemaking, temporary urbanism and others described in the literature (Carta et al. 2016; McGuirk 2014; Granata 2021).

Originating from the individual urban component, be it buildings, space, or infrastructure, the practices fit the characteristics of the urban regeneration intervention, as they are an easy way for the local community to use opportunities. Thanks to the use of ICT (Information and Communication Technologies) and social networks, these practices nurture more or less institutionalized relations within society and promote positive social innovation for the community (Coleman 1990). As already indicated in the introduction of this book, the scale of the project is not in this sense a determining and relevant factor for the purposes of the analysis, as this work, and urban reactivation, focus on questions concerning, instead, the ways in which trans-scalar social choices and implications mature (Bruzzese and Montedoro 2014).

Smith's contribution on the taxonomies of abandonment is mentioned, and in addition to analyzing the processes and policies of industrialization, de-industrialization and reuse of urban containers, it provides an interesting typological characterization of the artefacts subject to intervention.

A further interesting contribution for the purposes of this book is that of Di Giovanni (2018), whose description of the places of abandonment also considers voids, defining them as potential resources for redefining urban welfare apparatus, producing innovation and environmental sustainability, regenerating individual spaces, and reconnecting interrupted settlement patterns. These resources can, once activated, produce positive effects on neighboring contexts, reactivating relations between city spaces and social contexts. In addition to buildings that have become disused due to the loss (or reduction) of their value, meaning and usefulness, Di Giovanni also considers residual urban spaces, such as road junctions and intersections, urban fringes, uncultivated and abandoned spaces within the urban perimeter, plots that have never been built on, technical urban spaces, including quarries, landfills and various types of buildings. Spatial cases, as Di Giovanni (2018) himself defines them, which can result from neglect, destruction, abandonment, urban decay, but also from political and planning inattention or mistakes in the urbanization process. These causes most often originate from the disinterest of traditional operators but lend themselves to forms of informal sociality (Di Giovanni 2018; Franck and Stevens 2007) that participatory design, and particular combinations of urban recycling techniques, such as those mentioned above, can foster.

1.3 Taxonomy of the Built Environment Intervention

Whenever issues concerning changes within the city and the territory are discussed, a series of initiatives are tackled whose definitions begin with the prefix 're-' (Renewal, Redevelopment, Regeneration, Reuse, etc.). This essentially means that city transformation policies derive from a new way of thinking, which is implemented through processes of re-designing something that already exists (or existed in the past). These terms are often used as synonyms, and differences in meaning concern the indication of the needs and causes leading to the transformation of the territory, the policy adopted for the suggested solutions, the objectives and, in some instances, the types of intervention and the scope of application.

Another element that can be identified for the classification of intervention terms is the degree of obsolescence of urban, economic, social and physical functions, and their possible combinations, in order to re-design obsolete functions, transforming them into new functions that are compatible with the needs dictated by the changes taking place. Identifying specific terminology helps to pinpoint and describe different phenomena of urban transformation and to understand the cultural origin of the terms themselves.

Dalla Longa (2011) provides a specific taxonomy of interventions on the built environment according to their type and operational characterization. He identifies the following eight terms: Renewal, Redevelopment, Regeneration, Requalification, Recovery, Revitalization, Framework, Gentrification, Restructuring. The terms used refer both to the cultural context in which they are used, and to the nature of the phenomena of urban transformation. In this sense, the taxonomy proposed by Dalla Longa is the result of a difficult process of classification, both because some terms derive from a language that is not properly linked to the world of architecture and town planning, and because they sometimes assume different shades of meaning in relation to the characteristics of the transformation intervention, as well as to the morphological, political and cultural nature of the contexts.

In Europe, the following elements can be considered as generating differences in the use of terms related to urban transformation: the morphology of the city, political, institutional and regulatory variables. Some terms have influenced specific lines of intervention of the European Community, especially since Europe took steps in the 1990s to standardize different types of intervention within the union, such as the 'Towards an urban agenda' of 1997, the 'market efficiency' approach of the Urban 1 and 2 plans, the cohesion policy 2007/13, the environmental issues of the Leipzig Charter and the Toledo declaration, the smart strategy for sustainable and inclusive growth, and the more recent place-based approach, which introduced the need to integrate transversal administrative sectors related to human capital, social inclusion, energy and culture. This approach has been advanced through a number of policy tools with a budget of €351.8 billion—accounting for roughly one third of the EU's total budget between 2014 and 2020. The aim of the European Green Deal is to make Europe the first climate-neutral continent by 2050, and this could be an opportunity for an aesthetic regeneration of European cities, as stated by the New European Bauhaus.

The ensemble of these measures represents an important pool of methodological–instrumental references useful for supporting the urban reactivation intervention as a particular variation of the bottom-up regeneration intervention. For instance, the "Strategia Nazionale per le Aree Interne" (Strategy for Inner Areas) or SNAI, by promoting the integrated enhancement of both physical and social connective tissue through experiments aimed at community building, inspired the procedural model of reactivation intervention based on direct and participatory action by users. It is interesting to mention this strategy here because, right from the early stages of drafting the national strategy, different levels of government were involved, as well as private individuals, through various forms of association, with the aim of sparking forms of redemption and change that would feed on landscape and architecture. Widespread places of redemption, even minute ones, capable of creating economic and social development. These experiences have increased in recent years, composing a rich and multifaceted panorama of cases and practices, such as the InStabile project (Bologna), or Mercato di Lorenteggio (Milan), mentioned in this book.

1.4   A Particular Approach to Built Environment Intervention: Urban Reactivation

Over the years, the boundaries between the different categories of intervention described in the previous paragraph have gradually become blurred, both because there has been a partial overlap and fusion between the disciplines, and because the reading of the urban phenomenon as a sum of distinct and independent factors has disappeared. As stated above, the challenges posed by intervention on the built environment today call for the ability to work using integrated, multidisciplinary, and multi-stakeholder approaches, as well as to incorporate public and private interests, short-and long-term horizons, and different types of impacts (LAMA 2019).

Architectural and urban design is now increasingly linked to the themes of citizenship, environment and productivity, i.e., it does not rely on mere urban design, but also on strategic planning procedures in which social and production components collaborate on local territorial development. Intervention on the city does not only pose the challenge of modifying its components on a physical and formal level, but also the integrated management of its functioning mechanisms, starting with those which, through the concepts of sustainability and resilience, have a direct correlation with social issues. The concept of resilience, in particular, directly affects intervention on the built environment in the forms proposed by technological and environmental design when, as Martin and Sunley (2015, p. 15) observe, "a socio-economic system (the city) should be able to cope with adaptive changes in its economic structures and its social and institutional organization, in order to maintain or restore the previous development path, or even move towards a new path characterized by a more productive and efficient use of physical, human and environmental resources". To put it differently, there can be no ecology without social justice (Forum Disuguaglianze Diversità 2019), and this requires the ability to manage the project through horizontal alliances between organization and citizenship, typical of the collaborative approach practiced by technological and environmental planning.

In addition, issues of environmental quality, use of alternative energy sources, land consumption versus resilience and urban metabolism are becoming increasingly central to urban transformation projects (Amenta et al. 2022). Thus, projects involving tangible and intangible resources (economic/financial systems, users, environment/territory, energy) and their use are designed in an integrated manner, and for this reason the planning process must use new tools to manage the resulting complexity.

The term "Urban Reactivation" defines a mode of intervention on the built environment that combines the physical-spatial dimension of places with the economic, social and relational dimensions (Ostanel 2017; Venturini and Riva 2017; Venturini and Venegoni 2016; Fanzini et al. 2020). In this sense, the concept of reactivation underlines the act of putting something into operation, or putting something back into operation, whether it concerns the material components of a container (a building or a place) or the immaterial components of a content (a need, a function), starting from the exploitation of the opportunities that the architectural project contributes to unveiling, developing and accompanying. The definition is the result of research into the reuse of abandoned urban spaces in a selection of case studies (Fig. 1.1), that the New Generations Association, a European platform initiated by Gianpiero Venturini that analyses some of the most innovative emerging practices at the European level and creates a space for the exchange of knowledge and comparison between theory and production, carried out in 2015 with the support of the Creative Industries Fund NL, the Embassy of the Kingdom of the Netherlands in Italy and Fondazione Cariplo.

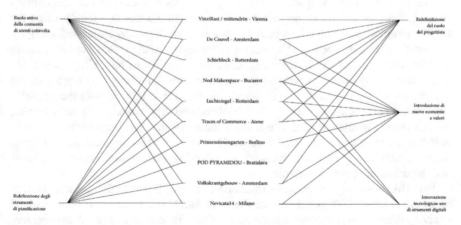

Fig. 1.1 Relation between a selection of case studies and the themes of the book. *Source* Venturini and Riva (2017)

Setting aside issues of scale, to focus instead on the relational dimension and processes (Bruzzese and Montedoro 2014; Ostanel 2017), in the analysis of the case studies we chose to place the large interventions side by side with the small ones, including the many initiatives triggered by unconventional parties, such as associations of private citizens motivated by ethical-social goals rather than economic ones. These are small interventions in perimeter areas but have the potential to trigger other improvement processes. These works was carried out in the spaces of the dense city that have already been urbanized, in affinity with the practices of regeneration and functional, productive and social revitalization that originate from the individual building and that the architectural project helps to reveal; light, delicate interventions that enhance and do not contrast with the existing building, "and it is in this sense that the ability of administrations and planners to listen to citizens becomes the key to the success and vitality of urban transformations" (Cangelli 2015; Fanzini et al. 2020).

Thanks to Venturini (2016) the term urban reactivation has entered the lexicon of intervention on the built environment "whenever we refer to an architectural project, indicating a more conscious use of the resources at our disposal". In this sense, the prefix 're' becomes crucial for understanding the nature of the approach: "when we talk about the re-activation of an urban space, we are, inevitably dealing with new models that currently lack connections to theoretical practice, and, as a response, a number of initiatives employing more experimental strategies and methodologies have resulted in, quite interestingly, a sort of systematization of the informal" (Venturini and Venegoni 2016). These are informal practices promoted by groups of citizens, associations, cooperatives or self-organized communities, which today can rely on ad hoc expertise to tackle more ambitious and structured objectives. In the Dalla Longa taxonomy scheme (2011) we place the term reactivation at the intersection between social reading and physical reading (Fig. 1.2).

The way in which the spaces are used by the users is decisive in all reactivation interventions, especially if the object of the intervention has an identity value for the community. The ways in which these assets are used, as expressed daily by citizens through their needs and habits, are of fundamental importance in determining the characteristics of the transformation process and its sustainability over time. The small empty areas, as well as the small unused buildings that have escaped the interest of traditional economic operators, stand side by side with the formally constituted and planned spaces of the consolidated city, constituting its informal, though largely animated, part. These small pieces of the city that have remained in oblivion, empty and abandoned, because of low economic interest, can generate new forms of community, and with them, trigger effective processes of reuse of buildings and the creation of new economies.

Participatory design becomes an important quality factor in the implementation of an urban project. The contemporary multicultural society, always connected and evolving thanks to technological innovations, demonstrates an increased involvement in the design of collective public spaces, characterized by impermanence and change. In this context, the needs of society can be interpreted through participatory design processes, which lead to the enhancement of the identity of the place as a consequence

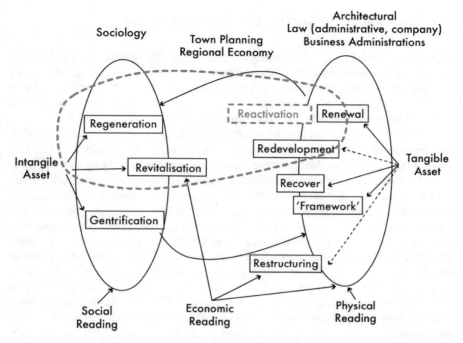

Fig. 1.2 Urban reactivation in relation to other definitions relating to the intervention on the built environment. *Source* Adapted from Dalla Longa (2011)

of the enhancement of the social networks that exist between the people who use that place (Manzini 2015).

Carta (2019) interprets these vibrant communities that care for the common good, fighting the crisis of values and experimenting with a new circular urban metabolism between the local and the global, as signs of a new Anthropocene (neo-Anthropocene). In this way, local communities become protagonists in the design of spaces for the community, in creating new visions of a future city with a view to enhancing the common good. In order to achieve better quality spaces, citizens come together in formal and informal communities to organize events and initiatives that also involve social inclusion. Urban reactivation, as an architectural practice, shifts its focus from the individual (common) good to the process of making it a useful resource to satisfy community needs in the sense that Perriccioli (2017, p. 27) stated: "An approach centered on a shared design that aims in the first instance to eliminate inequalities and social injustices, and which is opposed to the idea of a form of architecture that is totally subjugated to the rules of the market, expensive and confined to an ideological and stylistic self-referentiality" (Perriccioli 2017, p. 27).

1.5 The Social Perspective of Urban Reactivation

The realization that architectural design as a self-referential practice does not always produce high-quality, positive results for the common good has led in recent years to a change in the principles of both the culture of design and of the designer. The search for effective design solutions that have a genuine impact on the built environment is achieved by meeting people's primary, practical needs.

Operating from a social perspective means ensuring the active and collaborative involvement of citizens in strategic decisions, especially if these concern the city and the territory, increasing their awareness and civic sense, and creating new opportunities for them and their organizational structures. In this context, participation takes on a new meaning compared to the past: overcoming certain demagogic approaches, whose aim is to promote decisions that have already been taken, and others that are deliberately polemical, instrumental and preconceived helps to find the right mechanisms and the necessary resources to respond effectively and efficiently to the needs of the community that the 'normal' channels of intervention no longer manage to satisfy. If involved from the outset, people are less likely to criticize and more likely to direct their operational contribution to the reasons the project poses problems.

Understanding participation as an organizational form for shaping the relationship between policy and action in a process-oriented and collaborative way, using the project as a tool for responsible sharing of choices, is fully consistent with the objectives and principles of EU policies (Grote and Gbikpi 2002), which put the common interest before individual private interest and stress the importance of respectful, responsible and transparent action, promoting joint work and making the most of existing resources, also in terms of activating the capacities and energies that civil society is able to express and make available. The behaviors promoted by European policies is both pre-active and pro-active, anticipating the desired changes (Fanzini et al. 2018a, b) and then addressing them by involving institutions, citizens, associations and businesses, and is fostered in various ways:

– developing methodologies and tools (including telematics) to facilitate contact between the administrative system and citizens;
– mapping the resources in the area that could be involved in actions on strategic issues for the common interest and identifying meeting spaces for sharing choices.
– activating resources for participatory design on wide-ranging and diffuse issues of land construction and management, which can encourage creative collaboration between citizens and the administration and develop new skills for outstanding citizenship.

In all cases, involvement understood as co-responsibility is an indispensable premise for the adoption of shared public policies and participatory planning as a method for their effective implementation, the fundamental problem of which has always related to the forms of public involvement (Ostanel 2017). In this sense, the evolution of user involvement models, from participatory formulas to actual co-design, is changing the role of professional designers, as well as, of course, the users of the project, establishing new domains of creativity, which thanks to information technology can be extended and better directed towards the execution of the project. In this sense, digital tools could become an important instrument for managing strategies and selecting approaches. Pagani (2015) speaks of 'data' as an intangible structure of the city, which increases not only the basic knowledge of the elements, infrastructures and spaces of the city, but becomes, for companies, a potential element of innovation and exploration of the social and economic impacts of new regeneration processes. Similarly, Guallart (2015), using the example of Barcelona, describes Urban Habitat as the new project of fusion between urbanism and the environment derived from the networked society. It is a project that connects people, buildings, neighborhoods, cities, even states and the planet thanks to an information network that links all data and places in an overall project perspective. The citizens and organizations that participate and are involved in the project are fundamental to its success; all of them are interested in the common good, in a reactivated city that continues to be productive.

1.6   The Project-Based Approach of Urban Reactivation

In Sect. 1.1 it is stated that the economic crisis amplified the inability of public actors to manage common assets, both due to the reduction of the resources available and the decline of the ideals of equality and freedom that govern the delicate balance between different parts of the city. A further aspect that has amplified the aforementioned inability, is what Schiaffonati et al. (2011) attributes to the lack of a clear system architecture of the planning intervention and the excessive emphasis on regulatory aspects in place of more effective and efficient project quality control methods. Indeed, it should be noted, that while strategic and tactical interventions may trigger effective processes of renewal, there is a risk that they may turn into nothing more than temporary beautification or, even worse, into dangerous interventions of social estrangement or replacement when they are presented as impromptu exercises in aesthetic embellishment.

As Mussinelli (2018) has rightly pointed out, when urban intervention is unbalanced towards the purely ludic and fruitive dimension, it risks being reduced to a simple happening of events. The tactical dimension of the intervention, which could effectively guide strategy, risks being reduced to "a mere act of postponing ad libitum definitive and lasting interventions" (Mussinelli 2018). This is a far cry from Zanuso's concept of 're-appropriation of design' as a prospect of sharing and participation by users in the transformation of the environment, the territory and the city, and the ability to establish more appropriate relations with the environmental and landscape context. This perspective and capacity are based on the adequacy of the intervention in relation to the demand, and the consistency of the design choices with the necessary principles of utility, efficiency, durability and cost-effectiveness of the project (Schiaffonati et al. 2011). A perspective and a capacity, therefore, capable of producing effective and lasting results as they contribute to raising the level of knowledge, awareness and competence of citizenship (Mussinelli 2018).

Tactical orientation can, in this respect, legitimately precede strategy when the appropriation of contextual opportunities by local actors (Di Giovanni 2018) have the opportunity, through conscious and expert planning (Mussinelli 2018; Schiaffonati 2016), to relate to more complex synoptic frameworks of a cultural, political and phenomenological nature typical of the strategic dimension. The project is in itself a tool for innovation (Cetica 2003) through which it is possible to measure and control the quality of the intervention on the built environment. The project is therefore, as Schiaffonati (2016, 2017; Blundell Jones et al. 2005) states, the main tool for narrowing the gap between the inhabitant and the territory caused by approaches that are excessively unbalanced on the normative side, and for producing innovation.

The concept of social innovation refers to the modification of places according to new social structures; architecture becomes the mediating tool, which transposes social and economic demands onto the physical site. "The idea of architecture as the 'art of building for all', interpreted as a 'common good', is pursued, re-proposing its original function as an 'art of service' designed to satisfy the needs of a complex and multicultural user base, within a code connected to the rules of the community." (Faroldi 2017, p. 13). Professionals need to acquire new skills for the new design and action processes: social dialogue management, economic planning and management/maintenance skills. In this sense the architect assumes the role of mediator and facilitator in the implementation of projects, looking after the interests of the users involved. The role of the architect, besides being a mediator and facilitator for the attainment of the objectives of the implemented processes, is that of a social designer who helps those who inhabit their physical space; they prepare the structures where processes are coordinated in the perspective of co-design, where experts offer their design skills to those who work on the project idea (Perriccioli 2017).

The urban reactivation project originates from the community inhabitants and finds in the expert knowledge of the activating planner the possibility of relating to higher order interests such as the environment, social equity, and technological innovation (Mussinelli and Castaldo 2015). By mediating through the project between general strategies and instances of improved reactivation on a small scale, it is possible, in this context, to encourage the process of 'planning re-appropriation' expressed by Zanuso and by Mussinelli (2018) as a precondition for the practicability of bottom-up interventions. A possibility that is operationally supported by the maintenance of "a flexible structure that can change according to the needs of the embedded activities, constituting a dynamic masterplan that will gradually take root in the neighbourhood" (Belpoliti et al. 2015, p. 189).

References

Amenta L, Russo M, van Timmeren A (2022) Regenerative territories. Dimension of circularity for healthy metabolism. Springer, Cham

Belpoliti V, Boarin P, Davoli P, Marzot N (2015) Costruire nel costruito: Il riciclo urbano come strategia di rigenerazione sistemica del tessuto consolidato. TECHNE - J Technol Arch Environ 10:186–194

Blundell Jones P, Petrescu D, Till J (2005) Architecture and participation. Spon Press, London

Bruzzese A, Montedoro L (2014) Introduzione. Urban design: La via Italiana. In: Atelier 6. 'Atti Della XVII Conferenza Nazionale SIU, L'urbanistica italiana mel mondo', Milano, pp 753–755

Cangelli E (2015) Declinare la Rigenerazione. Approcci culturali e strategie applicate per la rinascita delle città. TECHNE - J Technol Arch Environ 10:59–66

Capasso D, Georgieff P (2015) Fare e Ri-Fare insieme lo spazio pubblico. La pratica di Atelier. Coloco dalla progressione intuitiva alla costruzione di strumenti adattivi per il progetto urbano. Abit Insieme 11

Carta M (2019) Vanguard cities of the Neoanthropocene. Econ Cult 3/2019:307–312

Carta M, Lino B, Ronsivalle D, Carta M (eds) (2016) Re_cycling urbanism: Visioni, paradigmi e progetti per la metamorfosi circolare. Listlab, Barcellona

Cetica PA (2003) La scelta di progettare: Paradigmi per una architettura della vita. Angelo Pontecorboli Editore, Firenze

Coleman JS (1990) Foundations of social theory. Harvard University Press, Cambridge

CRESME (2012) Città, mercato e rigenerazione 2012. Analisi di contesto per una nuova politica urbana. In: Working paper of the research "Riuso 2012: Casa e città per disegnare un futuro possibile", Milano

Dalla Longa R (2011) Urban models and public-private partnership. Springer, Berlin Heidelberg

Dansero E (1993) Dentro ai vuoti. Dismissione industriale e trasformazioni urbane a Torino. Edizioni Libreria Cortina, Torino

Ermacora T, Bullivant L (2016) Recoded city: co-creating urban futures. Routledge, London

Fanzini D, Rotaru I (2018a) Project anticipation as a tool for built environment social resilience. TECHNE - J Technol Arch Environ 15:101–107

Fanzini D, Rotaru I (2018b) Progetto e anticipazione della città futura. Sentieri Urbani, J Ist Naz Urban 28:53–58

Fanzini D, Rotaru I, Venturini G, De Cocinis A, Achille C, Tommasi C (2020) Collaborative reactivation of the built environment: a socio-cultural perspective. ACTIO-J Technol Des, Film, Arts Vis Commun 4:21–27

Faroldi E (2017) Architettura come materia sociale. TECHNE - J Technol Arch Environ 10:11–17

Forum Disuguaglianze Diversità (2019) Quindici proposte per la giustizia sociale ispirate al programma di azione di Antony Atkinson. https://www.forumdisuguaglianzediversita.org/pro poste-per-la-giustizia-sociale. Accessed 9 Sept. 2022

Franck K, Stevens Q (2007) Loose space: possibility and diversity in urban life. Routledge, London

Galdini R (2008) Reinventare la città: Strategie di rigenerazione urbana in Italia e in Germania. Franco Angeli, Milano

Di Giovanni A (2018) Urban voids as a resource for the design of contemporary public spaces. Planum, J Urban, Mag Sect 37(II/2018):1–28

Granata E (2021) Placemaker. Gli inventori dei luoghi che abiteremo. Einaudi, Torino

Grote JR, Gbikpi B (2002) Participatory governance. Political and societal implications. VS Verlag für Sozialwissenschaften, Wiesbaden

Guallart V (2015) Da pianificazione urbana a Habitat Urbano. TECHNE - J Technol Arch Environ 10:24–27

Henton D, Melville J, Walesh K (1997) Grassroots leaders for a new economy. How civic entrepreneurs are building prosperous communities. Jossey-Bass, San Francisco

Jones P, Evans J (2013) Urban regeneration in the UK: boom, bust and recovery (2°). SAGE, London

La Cecla F (2015) Contro l'urbanistica: La cultura delle città. Einaudi, Torino

LAMA (2019) Luogo comune. Progettare la rigenerazione urbana multistakeholder. https://age nzialama.eu/portfolio/come-progettare-la-rigenerazione-urbana-multistakeholder/. Accessed 18 June 2022

Manzini E (2015) Design, when everybody designs: an introduction to design for social innovation. The MIT Press, Cambridge

Manzini E (2018) Politiche del quotidiano. Edizioni di Comunità, Roma

Martin R, Sunley P (2015) On the notion of regional economic resilience: conceptualization and explanation. J Econ Geogr 15(1):1–42

McGuirk J (2014) Radical cities: across Latin America in search of a new architecture. Verso Books, London

Mussinelli E (2018) Il progetto ambientale dello spazio pubblico. Eco WebTown 8:13–20

Mussinelli E, Castaldo G (2015) Design and scale issues in the new metropolitan city: a study of the south-east homogeneous zone. TECHNE - J Technol Arch Environ 153–160

Ostanel E (2017) Spazi fuori dal comune: Rigenerare, includere, innovare. Franco Angeli, Milano

Pagani R (2015) Rigenerazione urbana e percorsi di innovazione. TECHNE - J Technol Arch Environ 10:11–15

Perriccioli M (2017) Social innovation and design culture. TECHNE - J Technol Arch Environ 14:25–31

Roberts P, Sykes H, Granger R (2017) Urban regeneration, 2nd edn. SAGE, London

Saccomani S (2016) Urban regeneration and crisis. In: EURA CONFERENCE 2016 City lights. Cities and citizens within/beyond/notwithstanding the crisis, Turin, June

Schiaffonati F (2016) The territory of infrastructures. TECHNE - J Technol Arch Environ 11:12–21

Schiaffonati F (2017) Per una centralità della figura dell'architetto. Eco Web Town 2(16):17–23

Schiaffonati F, Mussinelli E, Gambaro M (2011) Tecnologia dell'architettura per la progettazione ambientale. TECHNE - J Technol Arch Environ 1:48–53

Smets M (1990) Una tassonomia della deindustrializzazione. Rass Dell'architettura 42(2):8–13

Valente IPS (2017) Tassonomie dell'abbandono. In: Fabian L, Munarin S (eds) Re-Cycle Italy— Atlante. Lettera Ventidue Edizioni, Milano, pp 65–69

Venturini G, Riva R (2017) Innovative processes and management in the social reactivation and environmental regenerative project. TECHNE - J Technol Arch Environ 14:343–351

Venturini G, Venegoni C (2016) Re-Act: tools for urban re-activation, vol 1. Deleyva Editore, Monza

Chapter 2
Urban Reactivation

Abstract The chapter investigates the tools of urban reactivation, those supported by social innovation, meaning the use of forms of sharing and cooperation to valorize common goods and satisfy the common needs of society. For this reason, an important part of this chapter is devoted to the topic of participatory design for promoting a genuine collaborative involvement of the people. The outcome of the analysis of national and international case studies, as well as the results of the interviews and field experiments described in Chap. 3 of the book, contribute to qualify the methodology and the instruments of Urban Reactivation as well the role of the re-activator architect. This last figure is an expert capable of recognizing and guiding the reactivation intervention, taking on board the requests of the interlocutors and the possible contribution of the active parties involved. This is a cognitive enabling figure capable of acting as an agent of social, cultural and political mediation; a true urban curator who acts not—or not only—as a creator of forms, but as an enabler of processes and narratives.

2.1 The Urban Reactivation Approach, Linking Social and Physical Spatial Dimension

In order to be authentic and effective, the contemporary urban reactivation project must have the capacity to create a fair balance with the world in which one lives, i.e., it must explore and experiment with the limits of possible transformations that lead to changes in physical and social spaces (Tagliagambe 1998). The culture of design, in this sense, fits into the diversified system of relations generated by the changes taking place in recent years in the technological, economic and social spheres.

Within this system of relations, participatory design assumes a fundamental role: individuals are involved in the design of spaces, using them to find new balances and to take ownership of the places they inhabit. A collaborative link is established between society, the environment and individuals to tackle everyday needs, desires and problems together. Contemporary participative project culture, therefore, finds new connections between social innovation and design. Design becomes a tool for creating physical, service and social relationships between the local and the global:

the different facets of design form a resilient infrastructure that connects production and consumption of resources. The community rediscovers the advantages of living together to solve critical issues related to coexistence with the help of designers. It is in this interplay between expert design and diffuse design (Manzini 2015) that one can act on communities and local realities, responding to problems and bringing visible and non-visible changes to the territory. The knowledge acquired links the social-economic system with the technical-information system; the traditional institutional apparatuses interact with the social networks, defining new design processes. The relationships between top-down design logic, bottom-up forms and peer-to-peer forms give rise to new roles and processes for design disciplines, which necessarily become co-design. Design thus takes on a new meaning: it is the result of the combination of project culture, cooperation and widespread creativity to define sustainable transformation strategies (Losasso 2017; Manzini 2006). This new meaning also gives rise to new project actions with fresh objectives (strategic planning, placemaking, tactical activism, operational making); new relationships and new ways of involving the people in the project; new tools and approaches for disseminating, implementing and replicating individual and collective ideas and products (capability approach, scenario design, visual mapping, storytelling, visual representation).

Tools for direct and inclusive participation of citizens interested in the places to be reactivated, support the generation of long-term visions and the planning of future-oriented interventions, which do not remain on paper but become, in this way, more easily achievable. Experiences of reactivation initiated by citizenship to generate urban future visions without limits are made applicable with the support of technicians; collaborative forms of governance and the multi-directional design process resulting in the transformation of existing spaces from a social and economic point of view with the consequent enhancement of the existing heritage. Becoming a participant in change within a community of which one feels a member, because we are involved together in achieving common goals, not only increases social cohesion, but also leads to greater respect of the inhabitants for the environment in which they live. Architects, within this system, are the experts, the educated interlocutors of the inhabitants' needs, who can implement the process of physical and social change. At the same time the architect assumes the role of coordinator, connecting with the intelligence of the non-expert players of the project (Granata 2021) and is the transdisciplinary activator of the expert players (Di Battista 2006).

2.2   Flexible Methodologies for Urban Reactivation

The urban reactivation project cannot be considered as a top-down intervention without taking into account the characteristics of the local area and its communities. The initial premise for any reactivation intervention is therefore provided by the impossibility of applying predefined models without considering the many factors that constitute the richness of the elements represented by the local context and by the subjects that are part of it. We can therefore affirm that this is precisely one of the added values of the reactivation project: the element of uniqueness, defined by the stratification of information that we find in a territorial context made up of people, various entities and groups that represent the different souls of the territory; historical, economic and social conditions.

Although the application of a predefined model is not suitable for this type of intervention, this book aims to develop a flexible methodology that can be used as a reference base for managing the design path that characterizes a reactivation intervention. As mentioned, this model must be considered as an open tool, flexible, to be complemented with ad hoc solutions, and developed according to the many factors that define a place and the people who live in it. This kind of approach is part of the architect's DNA. Thanks to the education they receive at university, architects possess the ability and tools to play an important role within the urban reactivation processes, taking charge of management and anticipating solutions to problems that arise during the process. It is a central aspect that can be found in this book which emerged through the analysis of various case studies conducted by multidisciplinary groups often organized around the figure of an architect, managing processes on the one hand, and dealing with the intervention design, on the other.

As indicated by early studies on the crisis of current urban models developed in the 1960s, traditional top-down planning approaches have been shown to be no longer effective. This situation is countered by the vision of the mutating city which is continually self-analyzing by means of the representation and social sharing of the elements of transformation and which plans, in real time, possible solutions based on awareness and participation (Celaschi et al. 2020).

The years preceding the economic crisis were marked by a strong impetus in the construction sector, which favored the spread of a so-called top-down approach. This was an intervention model that, imposed from above by public or private bodies, did not require open discussion or any kind of negotiation with the community of users living in a given area. This category of approach generally fails to favor the underrepresented or economically weaker user sectors who are forced to accept the consequences of an intervention that, in many cases, pays more attention to economic parameters than to social ones. The top-down project depends on the sensitivity of a few people who make decisions based on a set of parameters that rarely consider the needs of local communities who are not given a chance to be involved in the decision-making processes.

Common, collective, shared, associative, participated are the adjectives that today better express the plural and varied dimension of the public space. Nowadays, reflecting on the destiny of such a space, often waiting for an identity, means questioning the meaning and mechanisms of the new 'actions' of the urban scene and recognizing the role that communication and collective participation play in them (Cirafici et al. 2015). The reactivation project arises as a reaction to the great global economic crisis of 2007–2008 after years characterized by major interventions that have favored the emergence of the so-called gentrification phenomenon, affecting various urban areas. It is precisely within these territories that we often see bottom-up interventions, bringing the human scale and the people who live in it back to the center of attention. With a more limited amount of public resources to invest than in the past, many areas have been left in the hands of active citizens and neighborhood associations who, during the hardest years of the crisis, have seen themselves abandoned by the institutions and had to take care personally of the management of these areas. Self-management, Do It Yourself |Do it Together, Participation and Collaboration, are terms that have come back in recent years, as a result of an increasingly widespread phenomenon within local communities active in their reference territory, and which, thanks to their direct involvement, have often been able to overcome the lack of public support. However, a few years later, when economic conditions appear to recover from the crisis, we once again begin to face a situation similar to that of the beginning of the millennium. We are in a post pandemic period that will presumably persist for a long time, especially among the more fragile and less wealthy sections of society, with once again a reduction of public resources to be invested in the territory.

In the last few years, the reactivation theme has gone from almost unknown to mainstream. Like all other themes, that in time become a sensation of the moment, reactivation has also been adopted by many local administrations as a mere slogan to announce their initiatives. The current public health crisis suddenly brings us back to the recent past. Currently, it is presumed that the next few years will be marked by a new crisis (economic, but also social), with a consequent reduction in the resources to invest in all sectors of society (Gehl 2020). It therefore becomes even more urgent to invest in what we have learned in recent years from the many experiences of urban reactivation that have successfully contributed to the recovery of the urban areas of various European cities. It is important to continue investigating this type of approach, offering new solutions and preparing the field for the many interventions that will be carried out in the years to come. The pandemic should lead us not to draw rash conclusions. Rather, we must recognize that a sensitive approach such as that of reactivation should maintain a central position in the contemporary debate for a long time as a responsible approach to address some of the urban challenges of the present and the future. An approach useful for all those involved in participatory projects of urban reactivation: on the one hand, the institutional actors, who could be called upon to review or update their policies, fostering participation, paying particular attention to the issues that characterize the needs of local communities; on the other hand, the same communities, directly or indirectly involved in the process, which are already playing an active role in the implementation of reactivation projects, and which, through this text, will be able to find useful ideas to apply to different contexts.

2.3 Urban Reactivation, from Temporary to Permanent

Many examples of urban reactivation begin as interventions of a temporary nature and then go on to consolidate through a series of successive phases: construction sites abandoned due to the economic crisis are transformed into opportunities to carry out recreational and/or social activities (Campo de la Cebada—Madrid); entire buildings realized in the period before the economic crisis of 2007–2008 (result of the economic boom of the late 1990s) remain unsold or unused and have become playgrounds where people experiment with new temporary functions, models of adaptive reuse (WTC II, Brussels—interview with Freek Persyn); interstitial spaces, parts of residual cities, in central or peripheral areas, seemingly without any value, become experimental fields for groups interested in their temporary reactivation, attracting new communities of users (Prinzessinnengarten, Berlin). Interventions that initially have a temporary and experimental character solidify over time and transform into neighborhoods with great social, economic and cultural potential. The construction site must be the place of a participated and shared transformation, ensuring that in the future vision of the urban setting, the transformation was not only the change of the physical area, but also how it is used (Cirafici et al. 2015).

It is useful to recall the concept of permanent temporariness, expressed by ZUS—Zones Urbaines Strategies, which followed one of the most impactful reactivation projects—the Schieblock + Luchtsingel. An example that arose in Rotterdam at the turn of the millennium and, during the years of the crisis, was reactivated through an open and inclusive process that is still ongoing today. A privately-owned office building was first occupied by a group of architects (ZUS) who, in a few months, obtained the commitment of the local administration to support the initiative, building the foundations for an intervention that is today positioned amongst some of the most effervescent cultural areas of the city.

The approach proposed by ZUS, defined as 'permanent temporality' is very useful in reactivation projects because it starts from the assumption that the characteristics of the intervention can undergo changes during construction. This is in part owing to factors that must be considered unpredictable a priori in times of economic, social, and political uncertainty, such as those we have experienced since the beginning of the millennium until the present. This is a central point on which this book intends to focus from the very beginning because it implies the need to think of flexible methodologies that can be adapted according to the changes that may arise during the realization of any project. Therefore, it is not a pre-established or scalable model, but rather a series of guidelines and tools useful for interpreting the different contexts.

The combination of these factors determines the possible success of a reactivation intervention, which must be managed by various experts, organized around a professional figure who, thanks to their background, is able to interpret the various instances that define a territory.

2.4 The Premise of the Urban Reactivation Intervention

The potential offered by the reactivation sites is almost never evident through a superficial analysis. For this reason, the reactivation territory escapes the interests of traditional operators "remaining there as scattered pieces of a pitiful monument to the inability to intervene on the city" (Clemente 2016).

Lenders, banks, private investors or public administrations often tend to make decisions that are not directed toward reactivating small parts of an abandoned or not very advantageous area at first sight. In reactivation projects, informal groups almost always emerge to express their interest in these areas without any support. However, many of these projects see their chances of success reduced precisely because they are often characterized by a high degree of improvisation. They feature limited (if any) economic resources, and institutional support is unreliable or even non-existent.

The question then becomes, what distinguishes the occupation of a public or private space, such as an illegally occupied building, or land transformed into an urban garden without the explicit support of the municipality, from an urban reactivation project with a solid chance of success? Even the former are projects that could potentially fall into the category of urban reactivation but often do not due to the lack of reliable support from the owners of the area or those who manage it. Over the years, before the economic crisis, we have seen numerous interventions, especially in the area of the illegal occupation of public or private spaces (Salvia et al. 2019). The development of these spaces, while demonstrating a high potential from a strategic point of view and their territorial impact, were stopped precisely because they were not set up with the right starting conditions. Building solid foundations therefore becomes a key point that defines the success of a reactivation intervention.

Urban reactivation projects often begin with great ambitions but in most cases feature weak institutional agreements. It turns into a successful intervention only when it has the potential to build a solid network of stakeholders that support/contribute/finance the initiative in the medium-to-long term. Without these premises, most interventions are doomed to fail. We have seen interesting cases, such as the project for the reuse of the Slovak Radio terrace in Bratislava which was interrupted because the owners of the space decided to transform it without considering the work done by the community of citizens involved during the first phase of the project. In just a few months, they transformed the project from inclusive and social into pure real estate intervention. In this project, the individuals in charge of the initiative had not entered into a clear agreement with the owner of the area who, once the market value had risen, transformed the intervention into a model far from the initial ambitions expressed by the community of local stakeholders. The starting principle of any reactivation intervention must therefore be the construction of a solid institutional network, as suggested by many of the experts involved in the research.

On the one hand, this book aims to insist on the uniqueness aspect represented by each reactivation intervention, and consequently the impossibility of defining a protocol or model applicable to any context, a priori. On the other hand, it advocates the importance of defining an open and adaptable strategy that can offer solutions to concrete problems in different contexts. It is believed that the uniqueness aspect of each intervention can be offset by the skills architects, as professional figures who, thanks to their backgrounds, become occult mediators able to identify solutions capable of responding to the critical issues that arise during the project. However, this book intends to clearly depart from the vision of the architect deus ex machina, who imposes ideas through a top-down model to promote the figure of the architect as an occult mediator of the reactivation intervention. A figure able to manage an articulated process knowing how to delegate to other experts, when necessary, in turn able to solve the typical problems of the reactivation intervention that the architect alone would not be able to cope with.

2.5 Urban Reactivation Methodology and Process

This paragraph aims to identify a starting methodology useful for tackling urban reactivation interventions, understood as a set of tools halfway between a model applicable to various intervention scales or various territorial contexts and one that focuses on the need to maintain open variable solutions according to the specificities that characterize the project and thereby an open model of permanent temporality.

The approach to reactivation indicated below is the result of four years of research which collected case study analyses, interviews with experts, and field experiences which, together, made it possible to extrapolate a series of guidelines that may be useful to those who set themselves the objective of undertaking a reactivation pathway in a given territory—a collective, a local authority, a cultural association, and so on.

Viewed in this way, the reactivation process (Fig. 2.1) can be divided into 4 steps: (1) Problem setting, (2) Partnership building, (3) Co-Design, (4) Consolidation of the intervention.

2.5.1 Problem Setting

The identification of the starting problem is the first of a series of steps to be followed to build a solid basis for the reactivation project and can in turn be separated into three points: (1.a) analysis of the conditions of the area and its inhabitants; (1.b) identification of supporting stakeholders; (1.c) definition of short, medium, and long-term objectives.

(1.a) The ability to analyze the territory can be considered as one of the skills of the architect, who is used to carrying out similar activities for the implementation of a variety of projects: consider an architecture competition, or the implementation of any intervention in the urban environment. The basis of any project is a qualitative and quantitative analysis to understand the conditions that characterize the context of the intervention. Physical/geographical (characteristics of the area where the project will be implemented), social (composition and characteristics of the users), historical/cultural or economic elements are, in general terms, those which must be understood by the person or team in charge of the analysis. Based on the characteristics of the intervention, we will analyze some aspects more closely than others: the composition of the resident population, when the community of people is a key aspect for the success of the project; the regulatory framework, to define the limits of a hypothetical intervention; the distribution of services in a neighborhood, to understand which ones may be the most suitable within a given social fabric. However, during the course of the intervention further features will emerge, which may or may not be considered useful depending on the project.

The tools for proceeding with the analysis of the context of reference are numerous and can be applied in both the analysis and user involvement phases.

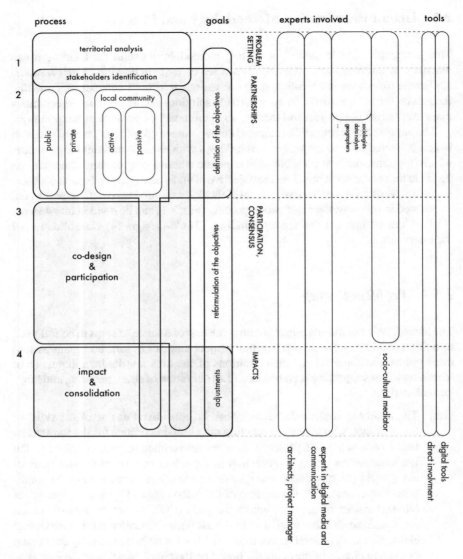

Fig. 2.1 Representative diagram of the procedural model of urban reactivation

These include active listening techniques, such as the neighborhood walk, which is useful for making an initial survey of the project area and understanding its characteristics first-hand. The direct involvement of the individual responsible for managing the reactivation project is, from the very beginning, one of the most important elements. Therefore, in addition to context analysis using the tools now available to everyone, such as the internet and

digital databases, which allow access to information directly from our worksta-
tions, context analysis must be conducted by combining digital and traditional
tools. Among these, in addition to the site visit, it is useful to interview (even
on an informal basis) people working in the area (neighborhood residents,
representatives of cultural associations, etc.). In this first phase it is necessary for the person in charge of analyzing the territory to be able to build a
personal toolbox, which will be useful for achieving objectives, justifying the
project choices and laying a solid foundation for the subsequent phases. A
central concept lies in the need to gain the trust of those who will be directly
involved during the reactivation process, which can only be achieved through
the involvement of a person who, over time, can be identified by the local
community as the reference expert for a given intervention. The reactivation
project has certain characteristics, in terms of the involvement of local stakeholders, which differentiate it from the regeneration or revitalization intervention: and this, among others, is one of the factors which is of great importance
during implementation.

The analysis of the area must therefore combine elements of a general
nature with others of a strictly local nature, which can only be discovered by
participating in the life of the neighborhood, through direct contact with the
people and gathering the elements necessary to understand the problems that
characterize the place of intervention. One factor to be considered is the level
of expertise of the people involved in the project: the people we will be dealing
with during the various phases of the project have a wide knowledge of their
localities. This factor suggests that the spatial analysis strategy should be based
on two aspects: (a) the need to be clearly structured from the outset, seeking to
gather information that meets the needs of the project, and the need to listen to
the implicit and explicit needs of the territory and local actors. This requires
the use of the various tools we have at our disposal, both digital and through
direct on-site participation; (b) the support of one or a group of mediators,
who operate at the scale of the intervention and are therefore familiar with
the dynamics of use of a neighborhood, or a given context. The local expert
in the role of mediator establishes an initial contact with the community of
inhabitants who will be involved during the intervention.

(1.b)   The outcome of this first phase of analysis of the conditions characterizing
the area and the individuals inhabiting it is therefore closely linked to the
second point of problem setting, namely the identification of the supporting
stakeholders. One of the most important aspects which must emerge during the
analysis is the network of actors operating at the scale of the intervention. This
is the only way to understand the dynamics of a neighborhood, the hierarchies
that have been created over time, the reference persons with whom it is necessary to deal in order to establish a direct link with all those who will be involved
during one or more phases of the project. As mentioned above, it is therefore
necessary for the person responsible for managing the project to be supported
by reference experts who can mediate, when necessary, convey the idea of a
project, present the project leaders to the different stakeholders who will be

directly affected by an intervention, and so on. Stakeholders operating at a local scale can be described according to three categories, which will be explored in more detail in the second point of the intervention strategy: (a) the local ecosystem formed by the people who inhabit the area; (b) bodies of a public nature; (c) bodies of a private nature. These three groups operate through hierarchies that have been consolidated over time and function according to their own rationale. It is therefore useful to investigate this issue before embarking on such a course.

(1.c) It is only after this initial phase of analysis that it will be possible to define a series of shared short, medium, and long-term goals that will respond to the specificities of an area. The definition of objectives must be thought of as a combination of several factors: like any project, in general terms, it is useful to set clear objectives which must be followed during the intervention. The reactivation project, however, must be able to adapt to certain needs or problems that may arise during the project, changing or redefining its objectives as it goes along.

Furthermore, it is assumed that the reactivation intervention is not a closed, circular process, which follows the dynamics of a classical architectural project (e.g., a building for housing with a public or private client, is commissioned, designed and built following predetermined standards and timeframes which must adapt to imposed factors, such as the regulatory framework, or the needs of a specific client). The reactivation intervention must be flexible and therefore able to adapt to possible unforeseen events which, as such, cannot always be predicted in the design phase. Adopting an open-ended approach therefore becomes a priority, building a clear strategy but with sufficient flexibility to adapt to different situations and contexts.

We will now see how, once the main characteristics of the place and the complex ecosystem with which we will be faced during the project have been identified, it will be possible to proceed with the construction of the network of stakeholders. However, it should be emphasized again that, out of necessity, the four phases are described in four separate chapters. It is not possible to define with certainty when a phase can be considered closed. On the contrary, it is useful to think of these phases as a set of tools to be used several times during the project, which, as noted above, should be interpreted as a flexible path to be viewed differently depending on the context. This will be followed by a discussion of the necessary agreements between the parties to be involved: the actions outlined in the next section cannot necessarily be handled in parallel with the problem setting, in order to understand, at an early stage, the margins of project implementation which, in most cases, depend on a solid construction of the project partnership.

2.5.2   *Partnership Building*

If we look at what is happening today in the field of built environment interventions, we can see that the relationships between designer, user, developer, and funder are not as well defined as in the past: roles alternate and, very often, partnership and collaboration agreements make all the difference. The impacts on the planning process are evident through co-creation, co-design, tangible collaborative tools and practices (Celaschi et al. 2020). In this sense, one of the aspects that needs particular attention is the identification of the stakeholders to be involved during the reactivation project. It is necessary to establish an ongoing dialogue with them to ensure a clear discussion at all stages.

Because of the flexible nature of the interventions, it is important that those involved are always aware of the decisions taken by the project leader, thus avoiding misunderstandings or potential problems that may arise when there is a lack of discussion. As outlined above, those operating within a given context can be described according to three categories: (a) the local ecosystem formed by the people who inhabit the area; (b) bodies of a public nature; (c) bodies of a private nature. The interests of each of these categories vary depending on the project.

Concerning point (a) the local ecosystem of people who inhabit the area can be composed of active or passive subjects. In the first case, it is a community of people/experts who organize themselves independently with the aim of undertaking a reactivation process. This community varies according to the project and is created through a variety of dynamics. In the InStabile project, Bologna, a group of people organized themselves with the aim of defining a shared path for the reactivation of an abandoned school. Through specific actions and a step-by-step process, the initial group grew, gaining the support of volunteers, neighborhood residents and associations, who contributed to the partial reactivation of the building. One of the critical aspects of the project was the lack of real ongoing institutional support and clear sources of funding: these two factors did not allow the project to develop according to long-term objectives, and there were various interruptions. A more successful case in this sense is that of the De Ceuvel—Amsterdam project: a community of active participants, in this case a group of young local entrepreneurs, came together under the guidance of Space&Matter—a local architectural firm experienced in urban regeneration projects, with the aim of reusing a contaminated area for the creation of a project that includes new workspaces, a research laboratory linked to sustainability issues, a hotel and a bar/restaurant. One of the substantial differences between the two projects, which indicates the positive outcome, is the presence of institutional support from the very beginning. De Ceuvel is in fact being built thanks to a call for tenders launched by the municipality of Amsterdam, which allocates funds for the realization of the intervention and an affordable rent for the use of the reactivation space for a period of 10 years. Secondly, the intervention is based on a sound economic model, which makes it more attractive to different types of bodies and institutions that might decide to participate or invest in the project, following clear rules. As we have said, the reactivation project, before becoming such, does not attract the attention of public or private bodies, but it has a high potential which, once made clear, makes it attractive again in the eyes of potentially interested parties. This aspect, which is of great importance, must be apparent from the problem setting phase. The territorial analysis must clearly explain the hidden potential of an area that has not yet attracted the interest of bodies keen to transform it, as is often the case, into an intervention of building speculation, or imposed from above without considering the needs of the local community. Explaining the added value of an abandoned territory is the task of the person or group in charge of the territorial analysis phase, which we can identify as the architect. The architect has the tools (those of the architectural project) to create a convincing narrative, to propose ideas and visions, and to anticipate scenarios.

The main beneficiaries of the intervention are the active subjects, those who are able to self-organize, understand the potential of an area and take action themselves to transform it. However, when successful, a reactivation project has the potential to extend its impact to the surrounding area. It is not aimed at a restricted community of people. Both in the case of the De Ceuvel project and InStabile, the reactivation of a contaminated area or a disused building becomes important for other local communities who benefit from the intervention, at first indirectly and, over time, directly.

One of the key elements when the reactivation project is carried out in this way, therefore, is the organizational form which is followed. A project such as De Ceuvel sees the Space&Matter studio as the leader, managing the various phases and consequently being recognized by all the components involved in the implementation of the project. It defines an implementation plan, communicates any decisions to the bodies involved, organizes the teams, updates the project plan and changes the objectives during implementation if necessary. There is therefore a clear hierarchy which gives each person, expert or body involved a clear position which must leave no room for doubt during the decision-making process in which they are involved. The definition of hierarchies differs according to the intervention: a call for tenders such as the one won by Space&Matter assigns the architectural firm the role of project leader. Consequently, Space&Matter manages the decision-making process together with the tendering body, which in the case of De Ceuvel was based on an open and participatory model, with extensive user involvement in the various stages of the project. In the case of InStabile, a local group takes the first steps, self-organizing. Once this process is completed, the foundations are laid for collaboration with other bodies. The negotiation process, in this case, is therefore in the hands of those who laid the foundations for the project, and can be managed according to different internal procedures, which must be clearly stated and shared at an early stage in order to avoid any conflicts that might slow down the implementation of the intervention.

When the ecosystem of stakeholders at the local scale is not the initiator of the reactivation project, the stakeholders can be defined as passive. The real aim here is to transform the community of individuals from passive to active, involving them directly in reactivation activities. A practical example is the project "Regenerating the Ciano", created in Piacenza in the framework of the call for proposals "Creative

Living Lab", First Edition. In this instance, the community of beneficiaries of the reactivation intervention were the inhabitants of the San Sepolcro working-class neighborhood, a residential complex composed of approximately sixty families of different nationalities and social backgrounds. The project saw the creation of a participatory process that produced the construction of a temporary pavilion within the communal garden area, little used by residents and dilapidated. Following a path planned over several steps, the inhabitants of the neighborhood were involved in a series of activities aimed at transforming the community from passive to active. The intervention can be seen as the first step in a more lasting urban reactivation process.

In the first of the two cases, when it comes to involving actors who are already active, it is clearly an advantageous situation, because it involves interacting with a community which, in addition to being familiar with the dynamics of the local area, already has its own internal organization and shared common objectives. In the second case, the reactivation project is more complex, precisely because it is necessary to build the basis for transforming the community from passive to active. This process is by no means a foregone conclusion: it must be planned in advance, even allowing for a high degree of uncertainty. When the community of inhabitants is not heterogeneous, many situations may arise which could make the reactivation project even more complex.

Concerning points (b) public bodies, and (c) private bodies—it is always useful to identify potential stakeholders, either public or private, who could play an active role in one or more phases of the project. It is possible to identify these subjects after defining the short, medium, and long-term objectives of the project, during the Problem Setting phase. The support of the public authority is fundamental from the outset, as repeatedly stated throughout the research—any intervention must have the agreement of the relevant municipality, ensuring its feasibility. In addition, depending on the scale of the project, the public authority may be involved at different levels. The most immediate one is simple approval for the implementation of a project, such as sponsorship. It ensures the ability to act in a given place, with an approved project plan at the very beginning and the granting of permits to carry out a given activity. There are other forms of involvement; depending on the scale of the intervention, a public body may provide funding, administrative support, offer subsidized user fees, or other forms of support. Each case has its own special features, and the relationship with the relevant local authority must be tailored to each case. In addition, the relationship with a municipality, or body responsible for the area, is more complex in certain contexts, such as in large urban areas where communication between municipal and project managers can be time-consuming and result in poor outcomes.

Generally, the involvement of private bodies is useful for two reasons: in the case of bodies such as foundations or banks, or private sponsors, they can provide financial support that ensures the success of the project, creating the conditions for following a project plan that develops thanks to concrete financing; moreover, being able to build a solid network of bodies guarantees greater prestige to the initiative. A private entity can also be the owner of an area to be reactivated: in that case, it is still necessary to collaborate with the public entity, creating a synergy that facilitates all phases of the project.

2.5.3   Co-Design

The co-design phase cannot be separated from the previous two phases. Through a broad process of exchange and participation, it is possible to define the project objectives shared by the parties involved and reshape them during construction if unforeseen events occur. In addition, regarding the co-design phase, different degrees of participation can be defined, but it is important to emphasize that in this case the architect or the responsible group of designers must be able to offer effective solutions that are consistent with what has emerged in the previous phases.

The methods of interaction between the various actors in the design process also evolve, bringing into play new forms of collaborative organization between users (networks, companies, associations, services) and new ways of involvement and participation of the inhabitants/designers, evaluating the quality and intensity of the interactions (Angelucci and Di Sivo 2013). The co-design phase, however, can take place in different ways, depending on the starting conditions of the intervention. Let us once again consider one of the successful cases included in this document, the De Ceuvel project, in Amsterdam. As a public competition announced by the municipality of Amsterdam, Space&Matter developed an initial proposal without defining the exact composition of the intervention, proposing instead an idea based on two levels: on the one hand, a clear proposal from the point of view of the functions to be included (neighborhood for entrepreneurship and start-ups) and the tools to be used (the reuse of abandoned boats to house these functions); on the other hand, another important aspect was the aim to build the project in collaboration with the future inhabitants. In this case, the co-design phase was perfectly set up by Space&Matter, who, without defining the formal architectural aspect, proposed the reuse of abandoned boats to be transformed in collaboration with the residents through a very clear planning process and, in a second phase, through participatory construction. As previously mentioned, De Ceuvel represents a particular case, because many specific optimal conditions were created in order to be able to conduct a reactivation project. In this intervention, the co-design phase can be described as consisting of a working methodology between the bottom-up and the top-down, also defined by VIC—Vivero de Iniciativas Ciudadanas, as 'middle-out'. A middle-out intervention ensures starting conditions, objectives and strategies are clearly defined from the first phase by an organizing body or by a client (in the case of the De Ceuvel, the Municipality of Amsterdam), which is in any case carried out through a process as shared as possible between all the parties involved, bringing together different sectors of society: citizens, the inhabitants of a specific neighborhood, public administration, private entrepreneurs, experts from different fields. Such case studies are rare. However, they have shown that they have an innovative project methodology, through a revival of the classic architecture competition format (with an auctioneer body, a starting loan, clear timing, a predetermined site to reactivate, etc.). At the same time, the format designates it as worthy of being carried out by decisions taken through the methodology of co-design and involvement of the community towards whom the project is directed. A similar process occurs in many European open calls that propose an implementation methodology such as that implemented in the De Ceuvel project, with the only difference being that in the case of a proposal financed through a European open call, a project location is not assigned. Rather it is the proposing body that is responsible for identifying a project area within which the proposal will be made.

Lastly, the intervention called Schieblock, carried out in Rotterdam by ZUS, is an example that realized the reactivation of an entire abandoned office building through the conversion of spaces, providing new functions useful for local communities. The methodology was similar to that of the two previous projects, differing only in that there was no auctioneer body. In this case, the same architectural firm, ZUS, supervised the management of the project, occupying the spaces of the building, negotiating subsidized usage fees with the Municipality while involving the communities that today use the building for various activities. These three examples show us that one of the pillars of the co-design phase, as also indicated in the two previous paragraphs (Problem setting and Partnership building), is the creation of a heterogeneous working group, able to represent all the actors involved within a particular territory, from public to private, including the groups directly affected by the intervention, and giving each of them a clear role. In the event that a particular reactivation project has not yet attained the level of the above-mentioned projects, the co-design phase can be used to test some ideas in the field using different methods. Let us once again take the Ciano project in Piacenza as an example. In this case, during the design and construction phase of the multifunctional pavilion, different meetings were held with the tenants of the social housing in the San Sepolcro district: a design workshop with the participation of students from the Politecnico di Milano—Polo di Piacenza, followed by another meeting with the tenants; the realization of the pavilion through a self-construction workshop with the participation of the students and some of the tenants of the neighborhood. This methodology can be applied to all reactivation projects that are in a first phase of study, proposing an intervention based on the testing of different solutions directly on the territory.

Upon analyzing some of the case studies, it may be concluded that the reactivation project can be carried out through either of the following ways:

(a) call for tenders managed by a public or private auctioneer body, which awards a loan for the reactivation of a project area, rewarding open and flexible design methods;
(b) self-promoted initiative, which starts from an idea developed by a group of experts, who aim to carry out a so-called 'unsolicited' project, without a real customer.

Between the two methods presented, the most common in the reactivation area is the (b) self-promoted initiative, which requires a greater effort in terms of planning, analysis and construction of the partnership but involves a high-risk rate which can be reduced by following some of the indications included in paragraphs—Problem Setting, and—Partnership building.

2.5.4 Consolidation of the Intervention

Once the design and implementation phase of the intervention is complete, the reactivation project can be considered concluded, but only in part. Many of the case studies that have been analyzed within this research show that a reactivation project continues along a new path of consolidation and changes that occur based on the feedback collected in the field. This is a key factor because it attests to the importance of the active involvement of the communities that take part in the project: whether it is a community of active residents, or of passive individuals indirectly involved, the reactivation project, when successful, demonstrates a strong social component, which translates into a united community moving towards the same objectives, constantly interested in improving the project with new interventions, even after the closure phase.

All the consolidated projects outlined above have followed this path: the De Ceuvel, where the community of inhabitants continue to develop the project area with the construction of new boats designed to host or improve the conditions of existing functions; the Schieblock, which enjoys an extremely united community of users in the local area, responsible for organizing the cultivation of a garden roof, the creation of cultural events in the adjacent areas, or in the cultural spaces on the first floor; the Prinzessinnengarten, an urban garden in Berlin that after more than 15 years of activity continues to reinvent itself to offer its community new functions, cultural events and services that attract an ever wider audience; MARES Madrid which, after the conclusion of the three-year period granted by the European UIA call, created an independent community of active citizens in the four reactivated buildings, who today manage the spaces independently, without the supervision of the authors of the project.

The consolidation phase of the intervention is therefore one of the most important because it highlights the significance of the work done in terms of participation. The community involved in the previous phases is the real protagonist here, as it matures to the stage where it can take over the independent management of its spaces, which is one of the most important benefits of the reactivation project. A space with a community of inhabitants who work in harmony, rooted in the territory, is a benefit that transcends the merely functional aspect of the intervention.

2.6 Instruments for the Practice of Urban Reactivation

Participatory design in the social sphere represents a methodological perspective that entails the collaboration of the various actors in a community (citizens or social groups targeted by an initiative, administrators and technicians) who work together on the design or implementation of a joint project, which will have positive effects on the communities participating in the process itself (Martini and Torti 2003).

Participatory design methodologies have their origins in two strands of research: the first contribution comes from Lewin, who in 1946 was the first theorist of participatory intervention research as an element of knowledge and therefore of transformation of what exists, where the objects studied become active subjects of the research; the second stems from social empowerment, theorized by Iscoe in 1984, which emphasizes the community as a subject capable—with its potential and resources—of bringing about constructive transformations in reality. Building on these two strands, a number of fundamental principles can be defined that shape and underpin participatory design methodologies: first, participation is a democratic process, where people decide on and control the changes that affect them; secondly, transformations decided by the community have a chance of lasting longer than those decided from above, since communities themselves, as resources of the territory, develop the capacities to deal with any problems that may arise without delegating to interventions from the outside; thirdly, no single actor intervenes, but several actors are involved in the intervention.

Participatory processes can be of two types: participation can be triggered by administrations and public bodies, in which case it is promoted from above (top-down), and the promoters, thanks to their external resources, initiate actions to consult and listen to citizens. The second type of participation process is bottom-up: in this instance, it is citizens who voice the needs of the area and, if they cannot find the necessary resources in the community, they ask the administration for support in implementing changes. This type of approach makes the participatory process no longer a mere top-down exercise in innovative techniques that is not reflected in reality but transforms it into a process in which the recipients themselves propose interventions and are therefore more interested in their implementation, because these interventions bring about positive effects on the territory and on themselves.

Through their participation, citizens share knowledge, information, needs, problems and visions; elements that are part of the project heritage. This creates a new sense of citizenship, a willingness of individuals to discuss their visions in order to align and outline intervention strategies that become the community's own. The different instruments of participation lead to the definition of these visions in the project. There are now a great many tools for involving citizens in decision-making processes concerning territorial transformations. There are more traditional tools, which facilitate communication, others that are more innovative and aim to build visions, projects for the future with the proposal of real operational strategies; some are more structured, with fewer participants and follow precise guidelines, while others seem to be unstructured, but it is the stakeholders themselves who determine their organization.

There may be several reasons to initiate inclusive decision-making (Bobbio 2004): when there are existing or potential serious conflicts; when support from others is needed to make/implement a decision, i.e., legal, or financial resources may be lacking, information, projects are integrated or policies are co-produced.

Participants in the inclusive decision-making process can take on different configurations: a first, simpler configuration may consist only of public institutions such as local territorial authorities or functional agencies; a second configuration sees the participation of organized groups present in the territory such as associations, sports and recreational groups, etc.; the third configuration sees the participation of non-organized citizens as well, in which case participation is voluntary and concerns restricted parts of the territory and should include most interests and points of view. The general principles common to the techniques are:—to help non-specialists understand;—to organize the decision-making process by providing a framework of shared rules such as stages, spaces, time, etc.;—to promote informality, and face-to-face discussion between participants;—to make all information available to all participants and to be transparent during the decision-making process;—to help interactions between ordinary people and experts.

There are, therefore, different degrees of participation in decision-making processes that precisely define the instruments of the participatory process (Bobbio 2004). The first consists only of listening to citizens, who remain passive subjects, and decisions remain in the hands of administrations and technicians. The second involves the participation of citizens in the choice of interventions proposed by technicians from outside the community. The third level sees citizens involved in the decisions of the administrations, which use these tools to initiate consultation processes to manage conflicts, pre/post-interventions, reach shared conclusions and promote deliberative processes. Places, opportunities and tools are needed to make participation a practice and, above all, to enable those who participate to maintain an active presence that counts (Brunod 2007).

Listening techniques are generally used in the preliminary phase of inclusive decision-making processes after identifying the agents involved (or more generally, the shareholders) once the issues to be discussed have been identified. A series of tools are available that can be used for passive as well as for active listening which can prove to be much more useful. Marianella Sclavi, an Italian sociologist, founder of Ascolto Attivo Sas, has disseminated this category of techniques in Italy in order to close the gap between groups of people who have difficulties understanding each other. The method of active listening applied to participatory processes is based on the concept of 'collective intelligence' expressed by the people, to which we can today match the concept of the 'connective intelligence' expressed by the architect. This idea is recognizable in moments of joint investigation and mutual learning wherein various parties engage with one another, co-creating and sharing new solutions that lead to mutual success and enjoyment.

The tools used by Ascolto Attivo for mutual learning and exploration of possible worlds come from decades of experimentation of international practice, modulated from time to time to be effective depending on the characteristics of each project and its needs. Some of the most frequently used active listening techniques employed in participatory processes are: the outreach, a practice of identifying people in order to consult and listen to their opinions; the territorial animation, similar to the outreach, it aims to raise awareness among local actors on the issues affecting the area to be transformed; the research-participatory action, a technique that takes advantage of citizens' local knowledge and then transforms it into actions that will contribute to change, improve or reactivate the area of intervention; the neighbourhood walk, during which participants communicate and share information, observations, experiences, memories, and questions that help in understanding the neighborhood; the listening points, elements located in the urban areas that are subject to transformation functioning as temporary desks to involve and listen to citizens; the focus groups are used to focus on a specific theme or phenomenon with the participation of a small group of people. The participants are selected appropriately such that they have homogeneous characters in order to not create disparities in communication that are too strong; the brainstorming, a technique that uses the dynamic of a game, in order to find possible solutions to a specific problem; the consensus building is a technique that allows participants to find solutions and make decisions in a polyphonic framework where all the points of view and identities of the people involved are respected.

The second level of participation is one that involves a greater degree of interaction. Among the techniques for constructive interaction, some aim to allow participants to build future scenarios, enabling them to look ahead and share their visions: such methods are described as building scenarios. Other methods help participants from different backgrounds, from the least to most skilled, to carry out complex reasoning, to think about problems and find possible solutions on a public platform: these are called simulation techniques. There are alternative methods which allow the participants to be autonomous and thereby free to define the themes and identify solutions by themselves.

The EASW (European Awareness Scenario Workshop) is an example of the building scenarios technique, launched in Denmark in 1994, it supports technological development as a tool to meet the social needs of a community. In practice, the method is developed with a workshop and elaborated in 3 phases. The first two phases are preparatory: scenario development and stakeholder mapping with the local organization. The third phase consists of a workshop with the aim of developing project ideas, supported by a facilitation team; The Action Planning is able to identify problems faced by the stakeholders on a daily basis and to propose solutions for a specific territorial context. Each participant presents to the group the positive and negative aspects of the neighborhood and then proposes possible changes, also taking into consideration the advantages and disadvantages; The Future Search Conference is a method of investigation to propose a feasible future that is either an improvement on the foreseeable one or completely different. It is developed in the form of a 'laboratory' and the participants, about 35–40, are asked over the course of two to three days to define common strategies for possible and desired future scenarios. The appreciative inquiry is a method that identifies, through a system of interviews or structured events, the most interesting and motivating fields for communities to implement processes of change.

Among the techniques based on simulation it is worth mentioning the design charrette, a participatory design process that involves citizens, experts, technicians, who activate an experience of knowledge exchange through workshops and roundtables; the 'planning for real' aims to help people, in particular the most introverted or least socially-skilled, to interact with each other in public as an alternative to public debates; the visual facilitation is a methodology of graphical recording on large sheets of paper during participatory planning meetings: the 'world café' technique employs discussion and comparison on issues of concern to the participants, divided in different tables and sessions of 20–30 min each.

The OST (Open Space Technology) is a technique founded on spontaneity, it is a creative laboratory technique based on the self-organization of groups, from 5 to 1000 people, who gather in workshops to discuss, work, and find solutions to complex issues; the neighborhood laboratory is a technique that involves citizens, through educational workshops, in the transformation processes related to their neighborhood. Various stakeholders, technicians, administrators, economic actors, organized and non-organized citizens, gather around in the presence of a facilitator. The methodology is flexible and proposes the creation of maps and diagrams of the participants' needs, resources, and future scenarios.

The following are some other techniques related to procedures implemented by public administrations that share decisions regarding territorial transformations with citizens: Participatory budgeting (Bilancio partecipativo) is a technique used by public administrations to decide, along with citizens, how to spend a certain portion of the administrative budget; the Débat Public, which emerged in France in the 1990s, is a technique used to facilitate dialogue between companies (interested in the realization of industrial and infrastructural projects) and the territory (the object of transformation projects). It is an instrument governed by French legislation and mandatory when a certain investment threshold is exceeded as well as for certain sensitive issues.

References

Angelucci F, Di Sivo M (2013) Resilience and quality of the built environment between vulnerabilities and new values. The role of technological planning. In: Society, integration, education, vol IV. Rezekne Higher Educ Inst-Rezeknes Augstskola, Rezekne, LV, pp 91–102

Bobbio L (2004) A più voci: Amministrazioni pubbliche, imprese, associazioni e cittadini nei processi decisionali inclusivi. Edizioni Scientifiche Italiane, Napoli

Brunod M (2007) Aspetti metodologici nella progettazione partecipata. Spunti 9/2007(Marzo2007-anno VIII n. 9):127–134

Celaschi F, Fanzini D, Formia EM (2020) Enabling technologies for continuous and interdependent design. In: Lauria M, Mussinelli E, Tucci F (eds) Producing project. Maggioli Editore, Sant Arcangelo di Romagna, pp 487–493

Cirafici A, Melchiorre L, Muzzillo F, Violano A (2015) Public space and contemporary city: the places of transformation. Hous Policies Urban Econ 2(1):65–86

Clemente C (2016) Marginali, dimenticati, dismessi. In: Clemente C, Baiani S (eds) B-side [Inserti urbani]. Il progetto tecnologico per la riqualificazione di spazi dimenticati. Edizioni Nuova Cultura, Roma, pp 13–22

Gehl J (2020) Public space, public life & COVID-19. https://covid19.gehlpeople.com/lockdown. Accessed 30 June 2022

Granata E (2021) Placemaker. Gli inventori dei luoghi che abiteremo. Einaudi, Torino

Losasso M (2017) Cultura tecnologica e dimensioni del sociale. TECHNE - J Technol Arch Environ 6–10

Manzini E (2006) Progettualità diffusa e innovazione sociale. Equilibri 3/2006:489–502

Manzini E (2015) Design, when everybody designs: an introduction to design for social innovation. The MIT Press, Cambridge

Martini ER, Torti A (2003) Fare lavoro di comunità: Riferimenti teorici e strumenti operativi. Carocci, Roma

Minati G (2006) La concezione sistemica. In: Di Battista, Giallocosta G, Minati G (eds) Architettura e approccio sistemico. Polimetrica, Milano

Salvia G, Morello E, Arcidiacono A (2019) Sharing cities shaping cities. MDPI, Basel

Tagliagambe S (1998) L'albero flessibile: La cultura della progettualità. Dunod-Masson, Milano

Chapter 3
Urban Reactivation in Practice

Abstract This chapter collects the outcome of an articulated program of activities whose results are based on an inductive-deductive research methodology. Interviews with experts, analysis of international case studies, video interviews and field work, have supported the material included in the previous chapters and provided the tools to understand the procedural and organizational model for urban reactivation through participatory practices and bottom-up urban transformation. Section 3.1 introduces an analysis of the results obtained through interviews with international experts from different disciplines through which it has been possible to isolate some features that anticipate the concept of urban reactivation. Section 3.2 refers to a few international case studies of urban reactivation providing a list of concepts to analyze and understand the process of Urban Reactivation. Section 3.3 presents a series of analytical focuses that investigate in detail specific aspects of Urban Reactivation such as the potential offered by digital tools, seen as an important resource that can facilitate participation. Section 3.4 presents the results of an urban reactivation project called 'Rigenerare il Ciano' (Reactivate Ciano), which took place in Piacenza in 2019–2020 in the working-class neighborhood of San Sepolcro and was awarded the first edition of the 'Creative Living Lab' by the Italian Ministry of Culture. The project presents the results of an urban reactivation process that lasted almost a year, in collaboration with local authorities and neighbourhood inhabitants.

3.1 Definition of Urban Reactivation from an Expert Point of View

The paragraph summarizes interviews conducted with ten experts from among those who were involved in the 2015 research *Re-Act: Tools for Urban Re-activation*: Jaap Draaisma—Director of the Urban Resort Foundation, Amsterdam; José Francisco Garcia Lopez—Former Director of Cultural Heritage and Urban Landscape, Municipality of Madrid; Jeroen Zuidgeest—Partner of the MVRDV architecture firm, Rotterdam; Dominika Belanská and Boglárka Ivanegová—Jedlé mesto Association, Bratislava; Marco Clausen—Director of the Prinzessinnengarten project, Berlin; Vivero de iniciativas ciudadanas—Research collective, Madrid; Paola Alfaro

D'Alençon—Researcher at the Technische Universitat of Berlin; Davide Dal Maso—Director of Avanzi, Sustainability by Shares, Milan; Marthijn Pool—Founding Partner of the Space&Matter studio, Amsterdam.

In the first interview, Jaap Draaisma—director of the Urban Resort Foundation11 in Amsterdam—presents the activities carried out in relation to the Volkskrant project. Draaisma refers several times to the need to identify from the beginning the users that best fit the type of intervention, in order to be able to lay the foundations for the success of the intervention. Draaisma adds that it is not possible to artificially build the group of interest and that it is therefore necessary to pay the utmost attention at this stage through a set of well-defined parameters that are sufficiently flexible to guarantee the outcome of the initiative. Finding a balance between these two factors is therefore an important element from the beginning.

José Francisco García Lopez—director of Cultural Heritage and Urban Landscape at Madrid City Council from 2013–2015, presents a reflection on the reactivation and reuse of abandoned buildings from an institutional perspective. The first major distinction introduced by the interviewee is that between public and private buildings, with the former being more likely to be reactivated as they are managed directly by the administration without having to go through lengthy negotiations with private owners. The legal framework of the Spanish capital is therefore central to define some basic points for the reactivation of abandoned buildings. Another aspect that emerges from the interview and that is common in urban reactivation projects, is the need for public support, a parameter that has a direct impact on the probability of a successful outcome. Many interventions that start from below (bottom-up) and envisage regeneration, reactivation of abandoned spaces for the creation of new functions, encounter great difficulties if they cannot count on the support of the local administration. If it is a large project, the difficulties are linked to the economic aspects related to the high rents or the investments necessary for the renovation and regulation of the building/space in question. Even in the case of smaller interventions, the support of the public administration makes it possible to simplify the procedures that define the different steps of the process, from the application for permits that guarantee the use of the space, to the confirmation of insurances, and not so much to facilitate the dialogue with businesses operating in the area. García Lopez argues that the reactivation project often starts from the bottom, without a real predefined client, but from the beginning it needs the support of the public body that can facilitate the process which could otherwise would be delayed by typical bureaucratic issues.

Jeroen Zuidgeest, partner of MVRDV, Rotterdam, presents his personal view on the topic of urban reactivation. Zuidgeest presents an initial reflection explaining the transition from a large-scale, i.e., top-down urban approach, which has been widespread since the 1990s, to an increasingly common approach that starts from the bottom up and takes into account the local specificities of an area, which vary from place to place. Among the different approaches mentioned by the interviewee, it is important to focus on the importance of introducing the use of digital technologies and social media within the design process, with the aim of involving the different stakeholders and to encourage their participation during all stages of the process. Lastly, reflecting on the role of the architect, who is directly involved in these processes,

Zuidgeest defines the architect as a 'poet', a creative figure capable of inspiring the various subjects involved with an anticipatory way of thinking. Architects often play a mediating role precisely because they are able to find the right balance between technical and personal skills, positioning themselves as experts capable of managing a project divided into phases and involving different stakeholders. In this sense, the architect has a central role, not only as a technician or designer, but also as an expert who can manage the complexity of the project.

Dominika Belanská and Boglárka Ivanegová led the Pod Pyramidou project in Bratislava and from the beginning proposed an approach based on the participation and involvement of the various stakeholders active in the neighbourhood. However, the project faced many problems, including the use of space owned by private individuals. After the first months of activity, which brought obvious results for the area and the communities involved in the process of gradual reactivation, the owner of the space did not renew the usage contract, blocking the continuation of the initiative (2016–2017). This is an important element to be taken into account in the reactivation processes that take root and develop naturally in the first phase of the 'occupation' of an area, but which in any case must be consolidated through solid foundations with all the partners of the project. The point made by Garcia Lopez regarding the importance of establishing solid partnerships with the public sector from the outset is therefore central to avoiding risks such as those that forced Pod Pyramidou to be abandoned.

Marco Clausen, director of Prinzessinnengarten, has led one of the most successful urban reactivation projects in Europe. The project follows an organizational model based on the participation and involvement of users of different kinds at each stage and on a sustainable economic model capable of producing capital that can be reinvested in the project itself (circular economy). Among the most important aspects is that of direct communication and involvement of users on the ground. Finally, a note on the learning process: Clausen focuses on the need for exchange from which we can learn and improve an approach that does not follow a model but rather adapts and considers the conditions of the local area. There are many points raised by Clausen which are often found in the other projects referred to during the interviews. In the case of the Princess Garden, Clausen indicates that some of the key points are fundamental to the success of one of the most established urban gardens in Europe.

VIC—Vivero de iniciativas ciudadanas, introduces the concept of city initiatives (iniciativas ciudadanas), informal groups that operate at the local territory level and are characterized by internal structures based on horizontal and shared hierarchies. The 'iniciativas ciudadanas' in Madrid work mainly at the suburban level, and all together are part of a network capable of generating an impact at the urban scale. During the interview, the Spanish collective focused on the need to rely on the existing network rooted in the local context. Every city has a local network of activities that becomes an added value to any participation-based project precisely because it is based on existing resources that, once activated and integrated into a system, add value to the success of the project. The question of the network is therefore interesting when applied to the urban reactivation project, which often bases its success on collaboration, exchange and multidisciplinary approaches.

The interview with Jan Jongert, co-founder of SuperUse Studios in Rotterdam, highlights aspects related to the concept of urban metabolism and the need to experiment with new economic models related to the construction and reuse of certain parts of the city. Another relevant aspect mentioned by the Dutch expert refers to the increasing use of sharing platforms, with the aim of facilitating the construction of new exchange and development networks. Online platforms not only facilitate participation processes, which are important in reactivation projects, but also generate an impact in terms of sharing and knowledge exchange. In a historical and social context where reactivation projects are increasing and it is not possible to apply prefabricated models for every urban context, it is useful to create a space for sharing, mapping, facilitating and identifying a set of methodological procedures.

Paola Alfaro D'Alençon presents the concept of Ephemeral Urbanism, developed in the framework of the academic research Ephemere Stadtentwicklung, conducted during 2012–2014 at the Technical University of Berlin12 (the concept is closely related to urban reactivation processes). In her research, D'Alençon identifies three types of projects that relate to the category of urban reactivation: those for which public administrations do not have funds; projects that are interrupted for various reasons; projects related to the creative industries sector. The second part of the interview focuses on the state of the profession and the role architects, who, according to D'Alençon, have lost their central role since the beginning of the economic crisis in 2008. On the other hand, architects have the skills to reposition themselves in this new context.

Davide Dal Maso offers his experience as an entrepreneur and expert in the field of urban regeneration, introducing the case of Avanzi—Sostenibilità per azioni, whose offices are located in a building that was restored after a period of inactivity and is now used as a co-working space and bar in Milan's Piola district. Part of the interview focuses on the complexity of urban reactivation projects, which often materialize thanks to the involvement of various actors who contribute during the project. In this context, Dal Maso insists on the need to define a new professional figure capable of managing the process, one who he believes should be part of the public administration. A figure of mediation between the public and the private, who facilitates a dialogue between the different parties involved in this process. Dal Maso identifies two keywords: dialogue and cooperation, which are crucial in the implementation of any reactivation action.

Marthijn Pool, one of the three founders of Space&Matter, an emerging architecture studio based in Amsterdam, introduces some of the work they have created, focusing on a few concepts that characterize the philosophy of the studio, having recently completed several interventions of reactivation. The interview presents a series of insights related to the definition of new economies associated with the realization of each project realized by the studio, where the community of residents is considered an added value in the design process. Pool explains the motto "We is the new Me" that characterizes many of the interventions carried out by Space&Matter.

Figure 3.1 provides a summary of the concepts which have emerged from the analysis of 10 interviews with experts involved in the Re-Act research: Tools for Urban Re-Activation (Venturini and Venegoni 2016).

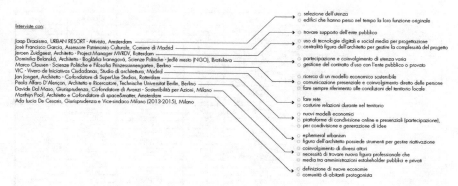

Fig. 3.1 Concepts which have emerged from the analysis of 10 interviews with experts involved in the Re-Act research: tools for urban re-activation (Venturini and Venegoni 2016)

3.2 Understanding Urban Reactivation

Several concepts were extracted from the analysis of the interviews included in the research '*Re-Act: Tools for Urban Re-Activation*': these concepts are useful when defining the research area more precisely and thus the concept of urban reactivation (Figs. 3.2 and 3.3). The concepts that emerged from this analysis point to a common approach among experts who, although from different fields, propose common visions and methodologies. Broad themes such as participation, the key role of the designer, sharing, and the involvement of large communities of users throughout all phases of the project become the starting point for the creation of an initial catalogue of 30 projects that relate to the broad category of urban regeneration.

The cases were selected through the definition of three parameters:

a. Keywords: the first parameter for the selection of the case studies is based on the concepts expressed by each interviewee summarized in the previous paragraph;
b. European network of emerging architects: the second parameter for the selection of the first list of case studies takes into consideration the analysis developed by the European network New Generations, uniting over 300 emerging firms from about 25 countries of the European Community. In 2017–18 New Generations conducted the survey 'ATLAS of Emerging practices: being an architect in the twenty-first century', which involved 95 emerging architectural firms completing an online survey, with the aim of understanding the changes that have affected the profession of the architect since the economic crisis. The results of this analysis show that the theme of reactivation is central in the activity developed by emerging European architects that during the research indicated projects which come under this category. Part of these projects have been included in this first selection of case studies;
c. Parameters: from the analysis of the data obtained from points (a) and (b), the following are a set of parameters which were taken into consideration when selecting the first 30 international case studies:—the territorial scale: referring

Initiative	Initiator	Typology	Scale	Permanent/Temporary	Public/Private	Key-words
1) De Ceuvel, Amsterdam	space&matter - architecture office	unused - contaminated area	medium	permanent	public	urban metabolism
2) Volkskrantgebouw - Amsterdam	Stichting Urban Resort - foundation	unused - office building	medium	permanent	private + public	artistic residence
3) Prinzessinnengarten - Berlino	Nomadic green - ONG	urban garden	medium	permanent	private + public	multifunction
4) Mercato Lorenteggio, Milano	Dynamoscopio Associazione - ONG	local market	medium	permanent	private + public	reuse
5) USE ME / Grindbakken, Gent	Sarah Melsens + Roberto Gigante	abandoned - docks	small	temporary	private + public	temporary installation
6) Schieblok + Luchtsingel, Rotterdam	ZUS - architecture office	unused - office building	medium	permanent	private + public	adaptive reuse
7) Nevicata 14, Milan	Pierluigi Salvadeo + Interstellar Raccoons	public square	medium	temporary	private + public	campaign
8) Matadero, Madrid	Ayuntamiento de Madrid - Municipality	abandoned - slaughterhouse	large	permanent	public	cultural center
9) POD PYRAMIDOU, Bratislava	Pod Pyramidou - ONG	roof terrace	small	temporary	private	unsolicited
10) InStabile, Bologna	Kilez Agency - architecture office	abandons - school	medium	temporary	private	adaptive reuse
11) Sweets, Amsterdam	space&matter - architecture office	unused buildings - mult. locations	large	permanent	private	adaptive reuse
12) Traces of Commerce, Atene	Traces of Commerce - collective	abandoned - commercial gallery	medium	temporary	private	temporary reactivation
13) Cinema Usera, Madrid	Todo por la praxi - architecture office	empy plot	small	temporary	private + public	itinerant installation
14) Piazza Gasparotto, Padova	LAB+ - collective of ONG	public square	medium	permanent	private + public	multi activities
15) M.A.R.E.S., Madrid	collective of public and private inst.	buildings	large	permanent	private + public	EU programme
16) NODmakerspace, Bucarest	Wolfhouse productions - architecture office	abandones - office building	medium	permanent	private	adaptive reuse, co-working
17) Pool is Cool, Bruxelles	Pool is Cool - ONG	multiple locations	large	temporary	private	campaign
18) Los Madriles, Madrid	collective of associations - architects	//	//	//	private + public	campaign, map
19) Tussen Ruimte, Amsterdam	Failed Architecture - architecture collective	multiple locations	small	temporary	private	workshop
20) Paisaje Tetuan, Madrid	Ayuntamiento de Madrid - Municipality	multiple locations in Tetuan	large	temporary	public	installations
21) Tropicana, Rotterdam	Rotterzwam - company	abandoned swimmingpool	medium	permanent	private + public	food
22) Made in MAGE, Milano	Ayuntamiento de Madrid - Municipality	multiple locations in Tetuan	large	temporary	public	installations
23) Largo Residências, Lisbon	Largo Residências - cooperative	abandoned building	medium	permanent	private + public	hostel and culture
24) Torre Oudaan, Anversa	collective of architects	abandoned - office building	large	//	private	campaign
25) Campo de la Cebada, Madrid	multidisciplinary collective	empty plot	medium	temporary	private + public	mixed functions
26) Cascina San Bernardo, Milano	società umanitaria	neglected building - farm	medium	permanent	private + public	food
27) Ciudad crea Ciudad	PKMN - architecture office	multiple location - Caceres	large	temporary	private + public	campaign
28) WTC, World Trade Center, Bruxelles	multidisciplinary collective	unused building - tower office	large	temporary	private	adaptive reuse
29) Tabacalera, Madrid	Ayuntamiento de Madrid - Municipality	unused building - Tobacco farm	large	permanent	public	museum
30) TransformCity!, Amsterdam	TransformCity!, company	multiple locations in Amstel 3	large	temporary	private	urban planning

Fig. 3.2 Thirty international case studies selected according to the parameters indicated in this chapter

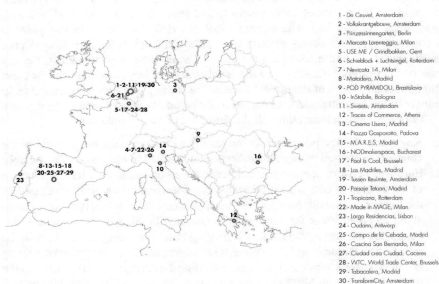

1 - De Ceuvel, Amsterdam
2 - Volkskrantgebouw, Amsterdam
3 - Prinzessinnengarten, Berlin
4 - Mercato Lorenteggio, Milan
5 - USE ME / Grindbakken, Gent
6 - Schieblock + Luchtsingel, Rotterdam
7 - Nevicata 14, Milan
8 - Matadero, Madrid
9 - POD PYRAMIDOU, Brastislava
10 - InStabile, Bologna
11 - Sweets, Amsterdam
12 - Traces of Commerce, Athens
13 - Cinema Usera, Madrid
14 - Piazza Gasparotto, Padova
15 - M.A.R.E.S, Madrid
16 - NODmakerspace, Bucharest
17 - Pool is Cool, Brussels
18 - Los Madriles, Madrid
19 - Tussen Reuimte, Amsterdam
20 - Paisaje Tetuan, Madrid
21 - Tropicana, Rotterdam
22 - Made in MAGE, Milan
23 - Largo Residencias, Lisbon
24 - Oudann, Antwerp
25 - Campo de la Cebada, Madrid
26 - Cascina San Bernardo, Milan
27 - Ciudad crea Ciudad, Caceres
28 - WTC, World Trade Center, Brussels
29 - Tabacalera, Madrid
30 - TransformCity, Amsterdam

Fig. 3.3 Map of the 30 international case studies selected according to the parameters indicated in this chapter

to the European context;—the time period: from 2004;—the period that slightly precedes the economic crisis, up to 2020, with projects still in progress;—the dimensional scale: the selection refers to cases of medium-small scale;—the state of conservation of the area: neglected or abandoned spaces, which have lost or modified their original function, or areas that propose a process of re-functionalization.

The case studies selected during the first part of the research were 30, as follows: De Ceuvel - Amsterdam, Volkskrantgebouw - Amsterdam, Prinzessinnengarten - Berlin, Mercato Lorenteggio - Milano, USE ME/Grindbakken - Gent, Schieblock + Luchtsingel - Rotterdam, Nevicata 14 - Milano, Matadero - Madrid, POD PYRA-MIDOU - Bratislava, InStabile - Bologna, Sweets - Amsterdam, Traces of Commerce - Atene, Cinema Usera - Madrid, Piazza Gasparotto - Padova, MARES - Madrid, NODmakerspace - Bucarest, Pool is Cool - Bruxelles, Los Madriles - Madrid, Tussen Ruimte - Amsterdam, Paisaje Tetuán - Madrid, Tropicana - Rotterdam, Made in MAGE - Milano, Largo Residencias - Lisbona, Torre Oudann - Anversa, Campo de la Cebada - Madrid, Cascina San Bernardo - Milano, Ciudad Crea Ciudad - Caceres, WTC | World Trade Center - Bruxelles, Tabacalera - Madrid, TransformCity! - Amsterdam.

The international case studies are spread over 9 EU countries, divided as follows: 8 in Spain, 7 in the Netherlands, 6 in Italy, 4 in Belgium, 1 in Romania, 1 in Greece, 1 in Slovakia, 1 in Germany, 1 in Portugal.

These are case studies that always fall into the category of urban regeneration, but not always into the sub-category of reactivation. Some, like Nevicata 14 (Milan), Los Madriles (Madrid), or TransformCity! (Amsterdam), are initiatives that have found space within this first selection due to certain specificities that needed to be taken into account for the continuation of the research: in the case of the Milan project Nevicata14 it is considered important to mention here the aspect related to the temporary reactivation of an urban space with major public involvement thanks to the use of digital media; the case study of 'Los Madriles' has been included in this first selection, although it is a mapping. In this case, the mapping of projects and initiatives located in the outskirts of Madrid has created the conditions to activate a network that in turn has an impact on the local territory and contributes to the regeneration of different disused places in Madrid. TransformCity! is the prototype of a digital platform that allows users to propose projects in Amsterdam's Amstel 3 district, promoting direct citizen participation through a digital tool.

The analysis of the case studies reveals a number of features that are considered in order to define the category of urban reactivation more precisely. There are two factors that are common to all the case studies:—the involvement of a large community of users on different levels of participation during different phases of the process (participation);—the application of bottom-up design approaches involving the need to search for new methodologies that transcend traditional planning tools.

These factors have been further analyzed during the research through a new series of 14 interviews with experts. The interviews, carried out at the fifth edition of the New Generations Festival, held at the Casa dell'Architettura in Rome from 24 to 26

September 2018, included the participation of 14 architects and experts from other disciplines with experience in the urban regeneration sector. Each interviewee was asked three questions:

(1) the first one dealt with different forms of governance of the public space;
(2) the second one investigated the impact generated by urban reactivation projects;
(3) the third one focused on the role of digital infrastructures.

With respect to the first question, the participants were asked to give their personal reflections on the management of public space in relation to the urban regeneration/reactivation project, with the aim of analyzing different approaches and methods of intervention practiced by experts who demonstrate extensive experience in this field. In the following pages, this concept is mentioned as 'Public space governance'. Regarding the second point, the research aims to understand the impact generated by urban reactivation interventions in different ways: economic impact, but also social impact, cohesion, as well as the impact in terms of interest generated towards the different categories to which the project refers. The third question focuses on the importance of digital media and is presented under the category 'New economies and values'; the research therefore focuses on digital infrastructures, and the potential offered by technological and mapping tools, as well as digital media and social networks, which are often used in reactivation projects.

Those involved in this round of interviews were: Emanuele Bompan—Geographer and Journalist (IT), Charlot Schans—Project leader New Europe—Pakhuis de Zwijger (NL), Ricardo Devesa—Editor-in-chief of Actar/urbanNext (ES), Joni Baboçi—Director for Urban Planning and Development of The Municipality of Tirana (AL), Simone Ierardi and Valentina Penna—Co-founders of CIRCOLO-A (IT), Annalaura Ciampi, Leonardo Tedeschi, Luca Vandini—Founders of kiez.agency (IT), Philippe Nathan—Founder of 2001 (LU), Mateusz Adamczyk, Agata Woźniczka—Founders of BudCud (PL), Ricardo Morais—Co-founder of Colectivo Warehouse (PT), Nasrin Mohiti Asli—Co-founder of Orizzontale (IT), Dario Tundo, Salvatore Peluso—Members of IRA-C collective (IT), Margherita del Grosso—Co-founder of Enter (IT), Bence Komlósi—Co-founder of Architecture for Refugees (CH).

The aim of these interviews was to further investigate some of the issues that emerged during the first phase of the research. Firstly, the interviews confirm what emerged regarding the two aspects mentioned above, common to all the selected case studies:—the participatory aspect, in the broad sense of the term, was discussed by almost all the interviewees as a central theme, highlighting the need to involve stakeholders in the different steps of the process;—the second point confirms the need to look at new tools and design approaches which transcend the traditional ones, with the aim of simplifying protocols and finding alternative legislative frameworks, thus reducing the timeframes and other investments.

This round of interviews also confirms the importance of investigating three new dimensions, which can be briefly described as:

- the redefinition of the role of the designer as an expert who gets involved in fundamental aspects of the process;
- the impact of these projects on the neighborhood and its inhabitants, which translates into new economies at the local micro-scale, and which are revealed through the appearance of new commercial activities, synergies between the inhabitants and local entities, or in the attraction of potential developers interested in investing in the territory;
- the use of digital communication tools, useful for managing/facilitating the process of involvement and active participation of the users.

These three elements were mentioned repeatedly in all 14 interviews, confirming their centrality in the research analysis. Most of the interviewees reflected on the redefinition of the role of the designer as an expert able to take on the complexity of these processes and mediate with the different stakeholders. The predominant theme is that of participation, which recurs in the interviews with varying interpretations. Charlot Schans (urban sociologist) focuses on the concept of placemaking, understood as a process that puts the community at its center. The collaborative aspect is therefore important for the co-creation of shared urban spaces. The application of these processes facilitates what Schans defines as a collaborative mindset. These are shared protocols that naturally become part of the way a community acts in the area. Joni Baboçi also focuses on a concept of great importance, 'balance', understood as the need to find a balance between the different actors that are part of the reactivation process. Baboçi mentions the architect as an active part of this process and points out the importance of creating a balance between all parties, through the search for new tools for dialogue and participation. Emanuele Bompan introduces the concept of 'power relation' arguing that at the beginning of the twenty-first century we witnessed a rupture in the balance of power, which has been reshaped with the advent of communication channels, social media and new technologies that amplify people's voices and thus play an important role in redefining decision-making processes. This often creates new conflicts, but at the same time people learn to deal with the conflicts by building a new way of acting on the territory. Ricardo Devesa focuses on the figure of the architect as an expert able to provoke a response, activate communities, mediate between citizens and design new uses of space. It is an important concept in relation to the theme of urban reactivation that places the figure of the architect at the center of the decision-making process, as an expert capable of managing the process. Bompan also talks about the concept of a circular economy as an element on which the values of the future are based, in order to build an economy focused on reuse. Although this concept applies to different areas of today's society, in the specific case of the theme of reactivation, it is applicable to the reuse of disused spaces, which translates into their transformation into new spaces for the community. Philippe Nathan, founder of 2001, talks about the participatory approach, as a tool applied in numerous projects realized by his company. The Luxembourg architect stresses the need to use the tool of participation only when it is really necessary, as it is often used only for communication and thus loses its potential. Another interesting topic presented by Nathan concerns the need to think of multifunctional buildings that can be used 24 h a day—and not only at certain times. Colectivo Warehouse focuses on two keywords: co-creation and hands-on approach, indicating two aspects that are fast becoming part of participatory planning. Co-creation is a concept that is not only related to the design of spaces, but also to the empowerment of people who, once involved, become an integral part of those spaces by taking care of them. The architect is able to manage this complex process and help the people involved to create a sense of belonging. Bençe Komlosi places the concept of inclusion at the heart of the discourse: The process of creation must be inclusive and open to all users directly and indirectly involved. To achieve this, communication channels such as Facebook and Whatsapp become important tools for maintaining contact and staying constantly updated. The role of the designer as mediator is also addressed by architect Margherita del Grosso, who refers to a 'negotiation of space': the presence of the architect in participatory planning processes is central to providing new food for thought, raising doubts and questions, mediating between the parties. These are skills that architects should have in their DNA. del Grosso also focuses on the concept of format: each project must be conceived with the appropriate format, and as far as urban reactivation projects are concerned, many could be conceived through participation formats of a temporary nature, such as design workshops involving residents or other types of event. The architect as a mediator of processes based on participatory practices is linked to the second concept that emerges from the analysis carried out in the research. The urban reactivation project has a direct impact on the scale of the neighborhood, generating microeconomics, synergies between residents and relationships that improve the conditions of the place. Many experts who participated in the interviews referred to this concept in conjunction with that of participation. IRA-C studio affirms the importance of creating appropriate conditions for dialogue and mediation as a necessary step for the production of collaborative and innovative architectural projects. The architect has the tools to perform this role. The tools for the implementation of communal projects are also important to create micro-economies: one of these tools is 'Municipal Crowdfunding'—which was used for the realization of the temporary reactivation project Gallab, in Milan. This concept is also mentioned by Joni Bacoçi, who argues that the participation of the different subjects is linked to the issue of finding new economies and sources of funding able to give sustainability to the reactivation project, which, according to Bacoçi, must involve the public and private sectors. Simone Ierardi and Valentina Penna, co-founders of CIRCOLO-A, focus on the economic potential and social impact generated by the reactivation of abandoned spaces, especially in Italy, where there are numerous spaces that fall into this category. By reusing buildings that have lost their original function, it is possible not only to contribute to the reuse and restoration of the building itself, but also to give new communities the opportunity to make their mark on the territory, with both economic and social benefits.

Lastly, the use of digital platforms was frequently referred to and constantly linked to the two previous themes (economies and role of the architect). The use of communication and dissemination channels, such as social networks, is increasingly widespread in projects that are based on the participation and extensive involvement of users of different types., The use of communication channels as a tool to facilitate new ways of involving users operating at the local level is increasingly prevalent; moreover, reactivation projects are becoming more widespread, starting from the creation of digital platforms that become a useful tool for fundraising, rather than for design. Annalaura Ciampi, co-founder of Kiez.agency oversaw the group that led the urban reactivation project InStabile, in Bologna. Despite being a group of architects, their concept of urban reactivation is closely linked to that of social reactivation. The reactivation then passes through the process, even before the reactivation of the artefact. The role of architects is therefore important as they are able to manage the two aspects: both that of the involvement of the various stakeholders, and that of improving the physical conditions of a space. In order to facilitate the civic engagement of the stakeholders, Ciampi highlighted that the use of social media and digital tools is becoming increasingly important, as they reach a broader audience that becomes central to these processes. Ricardo Devesa, editor in chief of Actar Publisher and UrbanNext also focuses on this concept by introducing the term 'digital arena', in relation to the theme of digital infrastructures: networks and platforms become a new space for sharing ideas, which act as a bridge between the virtual and physical world.

For the Portuguese collective Warehouse, the aspect linked to the use of communication technologies and media was central. On the one hand, the use of new technologies becomes essential for monitoring behavior in certain spaces; on the other, tools such as social communication channels, or crowd funding platforms like Kickstarter, become central to the involvement of the stakeholders at different scales of the project: from basic participation to fundraising. This concept is also vital for Orizzontale, who argue that the added value of these processes is not only verifiable from an economic point of view, but also in broader terms, such as that of multidisciplinarity. It is a concept that Orizzontale link to the theme of digital platforms, which in their opinion have the role of connectors. Thanks to the extensive use of digital media, it is in fact possible to stay connected, facilitate the exchange of information and, consequently, encourage the multidisciplinary nature of the project.

From the analysis of the interviews and case studies introduced in the previous pages, it is possible to identify a series of characteristics presented through the following 5 points:

1. interventions particularly carried out in conjunction with the 2008 crisis, which identified the need for new tools, practices, shared models;
2. areas which, for various reasons, have lost their original function and in which phenomena of environmental and/or social degradation are evident;
3. limited portions of territory that do not meet the requirements to attract the interest of potential public or private stakeholders;
4. direct involvement of citizens through co-design practices;
5. mainly urban contexts, without distinction between central or peripheral areas.

The research and analysis process has also highlighted three relevant aspects of urban reactivation that complete the definition of the previous five points and whose operational implications can be translated into new organizational and procedural design models: (a) the role of the architect as activator and coordinator of these processes, anticipating problems and opening new research paths, in order to encourage direct stakeholder participation through participatory and co-design practices; (b) the open and flexible nature of urban reactivation processes that favor the emergence of new economies at the micro-scale of the neighborhood, triggering innovative forms of participation, fostering inclusiveness and a sense of belonging of the communities involved through the different steps of the process; (c) the opportunity to exploit the potential of digital tools and new social media in order to facilitate the direct involvement of citizens and/or other categories. On the one hand, making the decision-making process more effective, clear and transparent and, on the other, as sharing tools that can strengthen the identity and sense of belonging to a community.

3.3   Urban Reactivation: International Case Studies

Starting from the definition that emerged from the analysis of the interviews carried out during the fifth edition of Festival New Generations (Rome, 25–26 September 2018—Casa dell'Architettura—Acquario Romano), it is possible to redefine the selection of case studies. The list of 30 case studies has been reduced to 4, which will be presented in this chapter through further analysis. The cases present the following characteristics:

– data relating to each project: construction period, scale, stakeholders, etc.
– analysis of the urban context: through the realization of territorial diagrams that indicate the characteristics of the place, of the intervention, of the project surface, and of the relationship with the urban context (center, suburbs, residential, commercial buildings, etc.);
– identification of the functions, which vary according to the context and users
– the users, based on the end users for whom the project was realized: a community of young entrepreneurs, the inhabitants, tourists, or other types of audience;
– objectives: which vary according to the type of project and are divided into sub-objectives (that may change according to the evolution of the intervention).

De Ceuvel

De Ceuvel consists of a plot of previously polluted land, transformed into an eco-space for creative and social enterprises in Amsterdam on a former shipyard along the Johan van Hasselt canal. In 2012, the land was secured for a 10-year lease from the Municipality of Amsterdam after a group of architects won a tender to turn the site into a regenerative urban oasis. The former industrial plot is home to a thriving community of entrepreneurs and artists, where all those involved have lent a hand to build Amsterdam's first circular office park. The plot hosts creative workspaces (inside 17 refurbished houseboats), a cultural venue, a sustainable café, spaces to rent, and a floating bed & breakfast. Old houseboats have been placed on heavily polluted soil, the workspaces have been fitted with clean technologies and it has all been connected by a winding jetty. De Ceuvel is a playground for sustainable technologies through experimentation. It is as energy self-sufficient as possible, processing its own waste in innovative ways (Figs. 3.4, 3.5, 3.6, 3.7 and 3.8).

Schieblock and Luchingel

Based on the idea of Permanent Temporality, this intervention introduces a new way of city making using the city's evolutionary character and existing forms as a starting point. When in 2011 it was announced that a planned development in Rotterdam, the Central District, had been cancelled, many existing office spaces were left vacant as a result. ZUS then decided to take matters into their own hands. They used a former office building, the Schieblock, to develop a city laboratory which currently acts as an important incubator for young entrepreneurs. With its ground floor called Depandance (local sustainable store, bar, culinary workshop, information centre), its co-working second floor and its rooftop field (DakAkker), Europe's first urban farming roof, it has become a prototype for sustainable development. Next followed the Delftsehof, a vibrant nightlife area, and then the child-friendly Pompenburg Park,

De Ceuvel
Korte Papaverweg, Amsterdam

 Public area, former shipyard
Amsterdam suburb

🕐 2012 - 2014

👥 **Architects**
Space&matter
Smeelearchitecture
Jeroen Apers architect
DELVA Landscape Architects

Experts and creatives
Metabolic *(clean technologies and sustainability plan)*
Studio Valkenier *(Design Café de Ceuvel)*

Social entrepreneurs
Woodies at Berlin and Brett Rousseau *(Contractors)*
Marcel van Wees *(project & general management)*
Waterloft *(financial advice)*

ACTIVITIES	BEGINNING	USERS	GOALS
creative workspaces cultural venue sustainable cafe floating bed&breakfast metabolic lab	winning of a public tender	community of entrepreneurs and artists	- circular economy - urban metabolism

Fig. 3.4 De Ceuvel, Amsterdam, project data

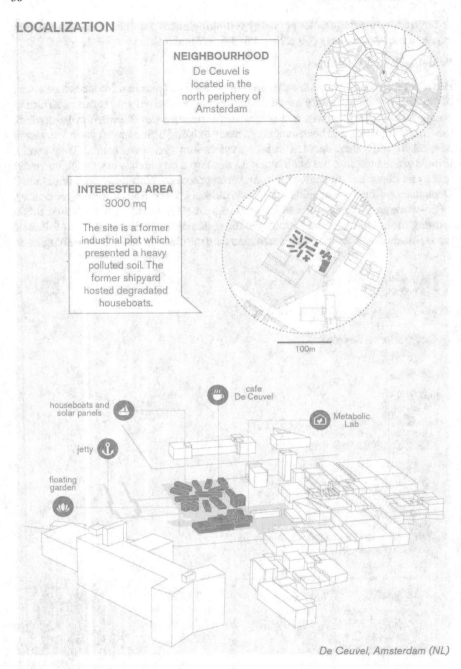

Fig. 3.5 De Ceuvel, Amsterdam, visual description of the project

Fig. 3.6 De Ceuvel, Amsterdam, aerial view of the project

Fig. 3.7 De Ceuvel, Amsterdam, aerial view of the project

with its vegetable garden next to the playground. These varied and new public spaces turn this former heart of Rotterdam back into being green and liveable, with the Luchtsingel bridge running throughout as a unifying factor. By simply increasing accessibility for pedestrians, the 400-m-long bridge will ensure synergy between

Fig. 3.8 De Ceuvel, Amsterdam, one of the phases of the realization of the project

the various sites. These distinctive connections give the area a unique position in Rotterdam's urban fabric (Figs. 3.9, 3.10, 3.11, 3.12 and 3.13).

INstabile

INstabile was born as an initiative towards the end of 2014 in the east suburban area of Bologna where a group of citizens spontaneously decided to recover the old civic centre which had been abandoned for over 30 years. The birth of new relationships among the inhabitants of the neighbourhood through the Social Street of the Portazza Village represented the push for a renewed interest in the building. For the original inhabitants of the district, with links to the building as the place of their studies and their childhood, seeing it closed or unoccupied and in a state of increasing decay had become a source of sadness; while for the more recent inhabitants, it represented a magnificent opportunity to make up for the need for a civic centre. This informal group of citizens started a participatory process together with experts through a 6-month co-design workshop. Over 200 citizens and 30 associations took part in the workshop, defining a recovery project and a model of how the building should work in the future (Community Creative Hub) (Figs. 3.14, 3.15, 3.16 and 3.17).

NOD Makerspace

NOD Makerspace is a creative workspace developed in the building of a former cotton factory, centrally located on the banks of the Dambovita river. In 2015, a group of multidisciplinary creatives set about transforming this abandoned industrial space into the first makerspace in Romania, so that the area was filled again with life and became a central hotspot for the creative industries in Bucharest. NOD is a dynamic

Schieblock & Luchtsingel
Schiekade, Rotterdam

◉ Public building, former office
Rotterdam city centre

◉ 2011 - ongoing

👥 **Architects**
ZUS Architects

Experts and creatives
CODUM *(turning vacant urban buildings into multi-company buildings for creative entrepreneurs)*

Public Institutions
Municipality of Rotterdam

ACTIVITIES	BEGINNING	USERS	GOALS
store	illegal occupation	- creative	- new enterprises
bar/restaurant	and consequent	community	facilitation
information centre	agreement for	- residents	- sustainable
culinary workshop	the concession		development
office spaces	of the building		- new urban
rooftopfield			connections

Fig. 3.9 Schieblock & Luchtsingel, Rotterdam, project data

ecosystem that aims to democratize design, engineering and creative education and welcomes designers, artists, engineers, inventors, freelancers and entrepreneurs. The space offers access to a wide range of tools and equipment for digital manufacturing and rapid prototyping. It consists of a co-working space, 21 private studios, one meeting room and workshop area (Fablab). On a larger scale, this place is a knot in an emerging network of creative industries in the city of Bucharest because it investigates experimental economic models driven by community groups (Figs. 3.18, 3.19, 3.20 and 3.21).

SWEETS

In 2012 the initiators of SWEETS hotel (a dedicated team of architects, designers, builders and artisans) presented a plan to the city of Amsterdam to transform the bridge houses into tiny hotel suites. The vision: to introduce travelers to new neighborhoods and unexpected experiences in the city. SWEETS hotel is indeed a one-of-a-kind hotel located across Amsterdam in 28 different bridge houses. In order to preserve the value of the individual buildings and that of the ensemble, each bridge house has been transformed into an independent hotel suite that accommodates a maximum of two guests, creating a distributed hotel that spreads from the upper north, across the IJ River, to the south of Amsterdam. The views over the canals and

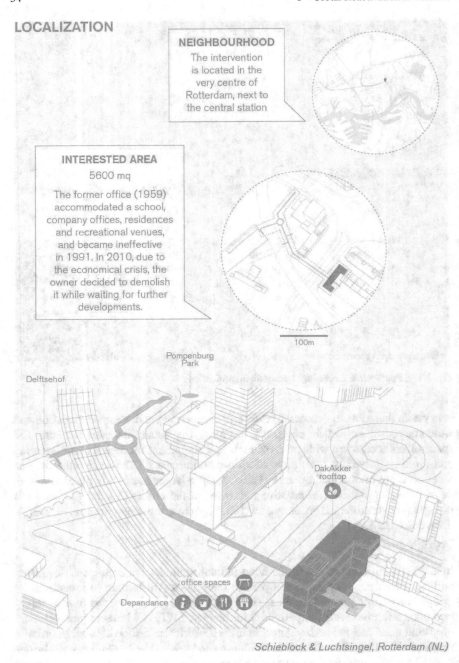

LOCALIZATION

NEIGHBOURHOOD
The intervention
is located in the
very centre of
Rotterdam, next to
the central station

INTERESTED AREA
5600 mq

The former office (1959)
accommodated a school,
company offices, residences
and recreational venues,
and became ineffective
in 1991. In 2010, due to
the economical crisis, the
owner decided to demolish
it while waiting for further
developments.

100m

Pompenburg
Park

Delftsehof

DakAkker
rooftop

office spaces

Depandance

Schieblock & Luchtsingel, Rotterdam (NL)

Fig. 3.10 Schieblock & Luchtsingel, Rotterdam, visual description of the project

Fig. 3.11 Schieblock and Luchtsingel, aerial view of the project

Fig. 3.12 Schieblock and Luchtsingel, view from the Luchtsingel bridge

the historical value of the bridge houses create a memorable visit. SWEETS hotel thus becomes a unique hospitality concept, one that has never been done before (Figs. 3.22, 3.23, 3.24 and 3.25).

Fig. 3.13 Schieblock and Luchtsingel, façade of the building

3.4 Specific Aspects of Urban Reactivation Through Emerging Architecture Practices

This section proposes three analytic focuses: (a) an interview with Freek Persyn, the architect and expert in adaptive reuse as well as founder of studio 51N4E; (b) an interview with the architects of Kiez.Agency, regarding the reactivation process that characterized the initial phases of the InStabile project; (c) an in-depth focus and analysis of the digital tools used by about 100 emerging European architecture practices.

(a) The interview with Freek Persyn analyses the concept of *Adaptive Re-use*, addressing the procedural and methodological question of the intervention, with particular attention to the aspect of temporariness. The Belgian architect began the conversion process of the World Trade Center in Brussels, installing his offices inside a building that had been in disuse for years. The temporary reactivation process took place in collaboration with other experts in the sector, who were personally involved, participating in the temporary reactivation of the office tower located in Brussels;

(b) The interview with Kiez.Agency proposes an analytical focus on the reactivation project InStabile, for the conversion of a disused school in Bologna. The focus includes an interview with the project authors and a detailed analysis of the phases that characterized the first stages of the project, paying particular attention to the procedural aspects, the involvement of the community residents and the financing process. The interview highlights the successful aspects of

INstabile
Via Pieve di Cadore, Bolonia

📍 Public building, former civic centre and elementary school
Bolonia suburb

🕐 2014 - 2018

👥 **Architects**
Architetti di Strada

Experts and creatives
ProMuovo *(citizen's association)*

Public Institutions
Municipality of Bologna
ACER *(Azienda Case Emilia Romagna)*

ACTIVITIES	BEGINNING	USERS	GOALS
civic centre urban square cultural events market square	self initiative of the residents	- neighbourhood residents - community of creatives	- succesful participatory project - new social dynamics

Fig. 3.14 INstabile, Bolonia, project data

the project, but also the difficulties encountered during the reactivation process, especially when it comes to finding economic support for its implementation. One of the aspects that often comes up during the reactivation project, which, as we have said several times, starts from scratch without the support of organizations or sponsors during the early stages, is that of the availability of economic resources;

(c) The last of the three analyses digital tools, providing an overview of the potential of digital media channels and tools and the ways in which they have influenced some aspects of the profession, paying particular attention to reactivation processes. The focus proposes a series of diagrams and interviews with experts analyzing how digital media channels such as Facebook, Instagram, Twitter or others can play an important role in managing the design process in general and, in particular, in engaging users, dissemination and public awareness in terms of reactivation intervention.

LOCALIZATION

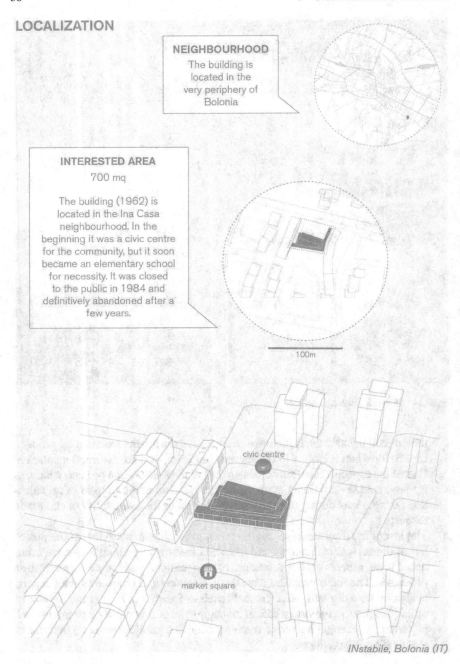

NEIGHBOURHOOD
The building is
located in the
very periphery of
Bolonia

INTERESTED AREA
700 mq

The building (1962) is
located in the Ina Casa
neighbourhood. In the
beginning it was a civic centre
for the community, but it soon
became an elementary school
for necessity. It was closed
to the public in 1984 and
definitively abandoned after a
few years.

100m

civic centre

market square

INstabile, Bolonia (IT)

Fig. 3.15 INstabile, visual description of the Bolonia project

Fig. 3.16 InStabile, area in front of the building

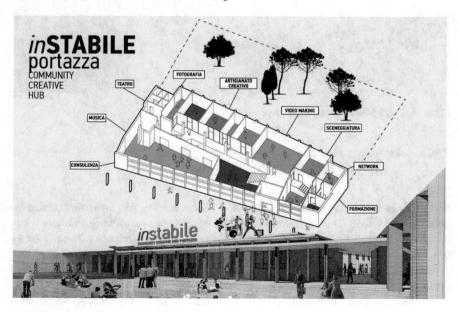

Fig. 3.17 Diagram showing the configuration of the spaces and their functions

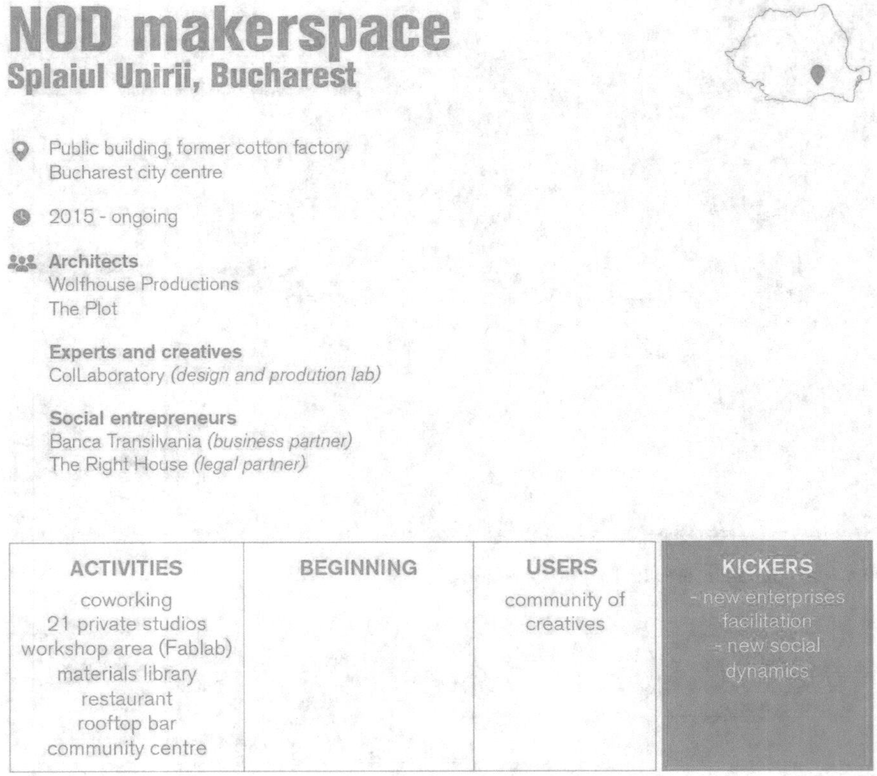

NOD makerspace
Splaiul Unirii, Bucharest

Public building, former cotton factory
Bucharest city centre

2015 - ongoing

Architects
Wolfhouse Productions
The Plot

Experts and creatives
ColLaboratory *(design and prodution lab)*

Social entrepreneurs
Banca Transilvania *(business partner)*
The Right House *(legal partner)*

ACTIVITIES	BEGINNING	USERS	KICKERS
coworking 21 private studios workshop area (Fablab) materials library restaurant rooftop bar community centre		community of creatives	- new enterprises facilitation - new social dynamics

Fig. 3.18 NOD Makerspace, project data

3.4.1 Adaptive Reuse—Interview with Freek Persyn of the 51N4E—WTC II, Bruxelles

51N4E is a self-steering collective and a collaboration platform that wants to empower people to be both autonomous and connected, using design to help overcome opposition and create integrated value and new experiences. It does so by organizing the supportive processes needed for a collaborative design culture.

Gianpiero Venturini (GV): *When you think of the important achievements that perhaps define these years, what milestones come to mind?*

Freek Persyn (FP): In the beginning it was really like a side office, which allowed us to do a lot of experimental projects. They were small, yet quite intense. Early on, we had the opportunity to start with the project called "Lamot", which was an adaptive re-use project, in which we didn't have the role of the architect. Rather our role was more like that of a scriptwriter, keeping the ambitions in line with the potential of the building. That was a key project early on. Another key moment in the practice was an opportunity to do, as a very young office, a very big project in Albania which

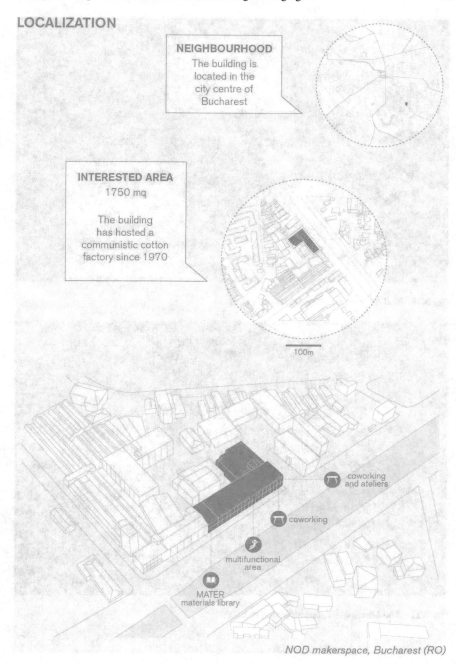

LOCALIZATION

NEIGHBOURHOOD
The building is located in the city centre of Bucharest

INTERESTED AREA
1750 mq

The building has hosted a communistic cotton factory since 1970

100m

coworking and ateliers

coworking

multifunctional area

MATER materials library

NOD makerspace, Bucharest (RO)

Fig. 3.19 NOD Makerspace, visual description of the project

Fig. 3.20 Nod Makerspace, coproduction spaces

Fig. 3.21 Nod Makerspace, coproduction spaces

SWEETS hotel, Amsterdam
Amsterdam

⦿ Public buildings, former bridge houses
Amsterdam city centre

🕐 2010 - ongoing

👥 **Architects**
Space&matter

Experts and creatives
Grayfield *(concept development and project management services)*

Social entrepreneurs
Suzanne Oxenaar, Otto Nan and Gerrit Groen
(founders of the Lloyd Hotel & Cultural Embassy)

ACTIVITIES	BEGINNING	USERS	GOALS
distributed hotel suites	self initiative of the architects	- local tourists - international tourists	- recovery of the urban heritage - sustainable tourism

Fig. 3.22 SWEETS hotel, Amsterdam, project data

was a young democracy. A third moment that was crucial, was when we started to really dive into more studies, prospections, strategies. We began taking the brief of a project very seriously and really research that. This really expanded the scope of the practice. Now, we do a very wide range of different types of projects.

GV: *What were the major themes or interests that you were preoccupied with at the time?*

FP: The interests, apart from how you shape and realize them, have always been the same. It related very much with the statement by Margaret Thatcher who said, "There's no such thing as a society". I think from the very beginning, we have been extremely convinced of that. For us, architecture is interesting if you design for that particular society, for how you can live together. We have always been interested in that question and we still are today. In the beginning, there was a lot of friction with a very neo-liberal approach which was quite strong in architecture. Today there is more ease in finding people that relate to our way of working. This has been there from the very beginning and it started from something very personal: how you feel you want to live your life. Now it has become more and more political because it is

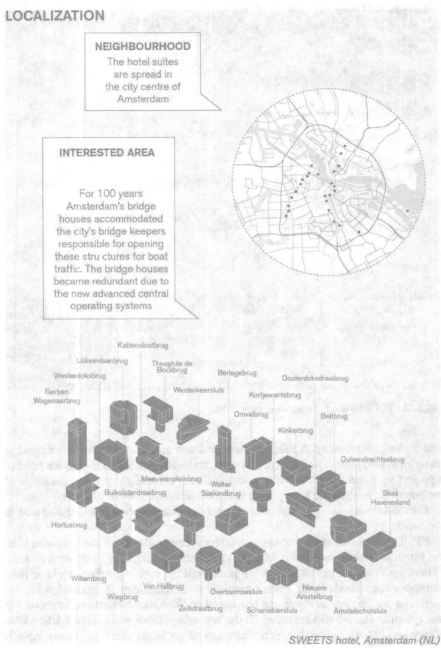

Fig. 3.23 SWEETS hotel, Amsterdam, visual description of the project

Fig. 3.24 Sweets, one of the 20 buildings reactivated in Amsterdam by space&matter

Fig. 3.25 Sweets, one of the 20 buildings reactivated in Amsterdam by space&matter

something you have to claim and it is something you have to fight for. It is something that today is still increasingly under pressure, this idea of 'society'.

GV: *Do you have examples of where, at the time, these ideas were translated into projects?*

FP: Going back to the very beginning, the first project we designed was a swimming pool for a private house. We didn't put it at the back of the house. Instead, it was the first thing you encountered when you entered the premises. The swimming pool became a place where interactions happen, and you were confronted with the act of going from public to private. That it is something that you always have to negotiate. It also comes with strong emotions as you would be undressed if you are using a swimming pool. This kind of creation of a conflict, but one that produces interaction, is something architecture can really work with. Architecture can trigger certain things to happen unscripted. It is a vastly different approach to create an architecture that produces these moments, and not just an architecture that fulfils functions. Therefore, the architecture that we do is very much about choosing where to place elements and how they meet other elements; not merely about how they look. The look is part of the atmosphere you build around it, but it is really a lot about spatial planning. This pool is a very good example of a thing that you see in a less explicit way in projects that we do later, but it is always these types of interactions that we look for and the way conflict actually produces spaces where people can interact.

GV: *How did these projects evolve into the projects you are working on today and how have the ideas evolved?*

FP: In the beginning, we were working on projects and trying to open up a project from the inside out. What we do today is rather to set up conditions so that the types of projects that we do, offer different project conditions. A good example is the place where we are. It is an existing tower called the World Trade Centre and it was built in the 70 s.[1] They evicted 13,000 people from this area in order to build this place. 40 years later, this place is 20–30% empty. Even the neighborhood is very vacant. We decided to move here and occupy it. This occupation has brought a lot of different dynamics into play. If you start to discuss the potential of doing another type of project here, different types of users could come into the neighborhood. It's less about working from within the project and more about creating the conditions around the project. If I talk of how things have changed, I think it is much more that we look for this dynamic to happen. Our practice became much more process-oriented and that is quite a strong component. We still create architecture, but we also organize things and we set up conditions for things to happen.

GV: *Could you describe the occupation process?*

FP: What we are doing here is not a counter-project but rather a type of alliance, in the sense that we work together with the people that own this building and we try to both align an agenda as well as develop a strategy on the go. It's clear that the potential of this area is huge and very central. It has a lot of capacity and a lot of space. In theory, it has a lot of the assets that we would describe as an excellent business

---

[1] The World Trade Centre where the interview took place, is in the Central Business District of Bruxelles.

district. In reality, it's difficult to feel those assets, so things have been organized and built in such a way that a lot of the qualities are not really there or are underdeveloped. Also, the neighborhood has been developed in a very one-directional way. Therefore, one way to rethink these areas is to think again, zoom out and think of a new vision. This is precisely what we are trying to avoid. I think we want to avoid creating a lot of distance; we want to come closer. In the closeness, we want to develop different solutions and different ways of doing things and different practices. The first step we took was also what made us meet the owners.

GV: *How would you describe the participatory elements of the project?*

FP: We were living in the city and we saw that this neighborhood was empty, and we proposed, in collaboration with the University of Hasselt, to do a workshop in the neighborhood with their students. That's where I met with Up4North, an alliance with eight main owners of this district. They were at the same time also busy setting up a possible approach to addressing these issues. We found each other at a moment that was incredibly lucky. From the workshop, we developed this whole dynamic. The workshop was set up with students, not in a totally internal world, but in an open way. There were public discussions and also discussions where developers were present. It was really the transparency that allowed the opportunity to create an alliance. The idea was of opening up, although the matters are still very fragile, where there are no answers, but there's still space to look for links, and for a shared agenda. What we are confronted with here is that on the one hand, there is the approach where you look for links, which is met again with a lot of resistance, because people cling to all of the mistakes that have been made and they hammer on these mistakes. Obviously, you cannot deny all these mistakes and injustices. The developers have played a noticeably big role in creating this injustice. However, we decided not to focus on this, and not to stand on the other side of the line while blaming them, but rather to look with a certain level of respect and engagement for where you can start to make links and start producing something which could lead to a better future. Over the course of a year, a lot of floors were filled. Architecture Workroom, from early on, joined the movement. There is an exhibition now that they organized, which actually is an on-site, work-in-progress exhibition, where these transformation projects, where you look for missing links and alliances, are actually produced at this location. It has become a spot where policy makers and citizens come. It's all still very fragile but, in this neighbourhood, it is a huge step.

GV: *Do you have a keyword to describe this mixture between the top-down and the bottom-up approach?*

FP: It is true that we need some good words to describe what is happening here. It's a weaving approach which tries to work on multiple levels at once. It's also trying to weave an alliance, so it is not that we start with a clear goal. The goal forms itself, just as how the alliance forms itself, just as how the mutual understanding develops on the go, just as how the number of people involved is also changing. Therefore, it's really an approach that tries to make a cross section through this bottom-up and top-down. Indeed, it is also focused on the idea that everyone in society has a role to play, and that we should try to make this connection happen. It's an approach that looks for connections, and through the connections, weave something that could potentially become very strong. In this process, there may be an attempt to avoid these big breaks or these big clashes. That doesn't mean that there's no conflict, nor does it mean that there's no disagreement, but there's a constant disagreement, and a constant search for agreement. I think that may be a huge difference. It's not about always stressing on the critical pushing aside of other people that have a wrong point of view. Somehow this process admits that everyone is partly wrong and partly right and that you should look for a shared purpose and develop on that.

GV: *How would you define your practice of today and how the practice is organized?*

FP: It's clear that we approach architecture from an idea of the built environment, and not from an idea of products. It's not that you have a question, and then we make the product that matches those questions. We think in a more contextual way. Context, however, is not about how buildings can fit in, but rather about reshaping the context itself. I think that this idea is something which we encounter all of the time; that we live in an inherited environment, we want to change that environment, we want to change how we live in it, so we have to work both on these new practices and these new environments. We also must work in a way that these environments will have activities that are constantly changing. It becomes about re-using but in a way that the use of a building is something that should be open-ended. We work a lot on this interplay between structures, adapted structures, and adaptive structures. We try to design possible new business models, and possible new briefs that you invent for the architectural environment. If you then look at the way the office is structured, obviously we still have a lot of people working in the office on the traditional tasks of an architect, to develop building details, quantity bills, and designs. Design allows you to see the synthesis of things. If you sit with people around a table from vastly different corners of the world or disciplines and design, you can discuss in a different way. We use design to do that: to work in situations where you look for new possibilities, putting together different perspectives. This working on transformation has allowed us to develop a new attitude, a new way of working, to design and dialogue, and to use design to facilitate that dialogue and to really go a lot faster in finding solutions. Recently, we have also started to focus more on the natural aspects of the built environment. Not simply working on reusing buildings, but also partly taking away buildings, and taking away hardscape; to bring in a different climate and quality by creating buildings that are healthier. You need plants to do so, on many different scales.

GV: *What do you think is the role of the architect? How do you square the disconnection architects are perceived to have from the rest of society?*

FP: What we are trying to do is to develop the architect less as the expert or the author, in his own corner trying to invent a solution, but rather as someone who has remarkably interesting tools to mediate different people. These tools actually allow us to do this in a very productive way. The fun of working on something, is that you can actually invite people to step into that. That it is the reason we enjoy doing architecture. The mediating role, or perhaps, the relational role of an architect is something we believe in very strongly. The fact that we have developed so much in a studio that is still not so old has to do with this attitude.

GV: *What are some of the key topics that you think will become increasingly relevant in the future?*

FP: We are extremely interested in transformation and how transformation goes through experiences that you build. We like to explore how the way of doing architecture is no longer from a written brief to a built result, but rather changing a built result into something new through temporary occupation. We want to have a kind of openness in how you define what the program for a building is. As a result, the building becomes spaces with more margins. That's where transformation is interesting because a transformed building is always a building that is unadapted. This lack of adaptation brings a lot of possibilities to find new ways to define how things relate to each other. It's not written out first, and then built. Instead, it's like reshaping something and finding new opportunities in that. When you look at the way in Europe, you see how we have built almost to the brink of suffocation. Thus, to open up to work in that built mass and change it and invent new ways of using it, is something we believe in very strongly.

GV: *What is the added value of temporary occupation—like the concept of ZUS of permanent temporality?*

FP: What is very striking is that when you invite people to think about something, they start to categorize, and a lot of things become really impossible. When you invite people to do certain things, your experience is much richer, and you can slide into new solutions more easily. The experience allows a lot of contradictions to simply disappear and that's what I think the biggest advantage of building through experience is. The range and resolution of your discussion is so much wider. It has become a big tool in urban planning. For instance, thinking about bike lanes, if you install them in a temporary way, people adapt them to their uses and established patterns. If you discuss things with them in a meeting room, around a table, the discussion is closed. There is a very big margin in reality, and that's I think very interesting.

3.4.2 Neighbourhood Economies—Interview with Annalaura Ciampi—InStabile, Bologna

The InStabile building is located in Bologna, via Pieve di Cadore 3, and was inaugurated in 1962 as a Civic Center inside the INA-casa residential complex in the Villaggio Portazza district. It kept this function for the first 2–3 years, then by necessity it was turned into a primary school. It was closed to the public in 1984 and used for a few years as a national archive, before being definitively abandoned.

Starting from 2014, the birth of new relationships among the inhabitants of the neighborhood through the *Social Street* of *Villaggio Portazza* represented the push for a renewed interest in the building. For the original inhabitants of the district, with links to the edifice as the place of their studies and their childhood, seeing it closed or unoccupied and in a state of increasing decay had become a source of sadness; while for the more recent inhabitants it could represent a magnificent opportunity to make up for the need fora civic center. Among the promoters, many were already active around the *Pro.Muovo Association*,[2] who the inhabitants subsequently met in the context of a 'street party', and who together with them became the hard core of the redevelopment project.

Realising their limitations in carrying out the process, the promoters looked for expert technical figures, represented in the beginning by the *Architetti di Strada Association*[3] and subsequently by *kiez.agency* which could help them to understand how to reopen the property and also be engaged in the process.

kiez is an agency that explores the traditional boundaries of architecture and urbanism activating process-based and socially sustainable transformations. They design and manage processes of urban space and territorial improvement investigating the public, private and common demands and promoting initiatives able to strengthen hidden resources, tie spatial and social relationships and turn urban space into places of opportunity. Kiez has been involved, since its creation, in the "INstabile—Community Creative Hub" process. In 2016 kiez.agency was awarded by Mibact and CNAPPC the design project for a new community center in the national contest for the regeneration of peripheral areas.

Gianpiero Venturini (GP): *Could you please outline the origins of the project?*

Annalaura Ciampi (AC): The INstabile project was officially conceived in 2015, when we participated in *Coop Adriatica*, a call for tenders that won us a €10,000 loan repaid in 4 installments over 2 years. Although weak, that was the spark that made us start investing energy in a process of citizen involvement: from the initial 10 inhabitants who came to us we managed to involve about 200 people, through a shared laboratory and a street party, proving that the building could be a communal and public space once again. During the participatory process we defined that what was missing in the residential district were cultural events. For this reason, we have defined the model of the Community Creative Hub, which was exactly what the building could offer once again. This type of cultural model aims to create a circle of exchanges between two different communities: the local community and a community of entrepreneurs and creatives with the desire to invest time and energy to make this cultural project work.

Once the planning phase was over (and the first loans ended), we had to deal with the renovation works on the building that we had not managed yet. In the absence of economic resources, we decided to enhance the human and relational resources we had, in order to share knowledge and skills that could make the project grow. That's why we created an abacus of skills by 'listing' participants according to their current job, 3 things they could teach and 3 things they wanted to learn. At the same time, we invited other figures who had already carried out similar projects to demonstrate to the inhabitants that the process, though difficult, was feasible.

---

[2] Associazione Pro.Muovo is a non-profit-making association with social and cultural aims, created to foster opportunities for creative and personal development, cultural meetings and socializing.

[3] *Associazione Architetti di Strada* aims to improve the answers to social and housing issues through sustainable projects. It consists of architects, engineers, urban planners, experts in human rights, communication, participation, energy and environmental sustainability.

To carry out the works on the property we contacted the owner, ACER (Azienda Case Emilia-Romagna)[4] and the Municipality of Bologna, with whom we started a negotiation that allowed us to manage part of the property (180 square meters out of a total of 700) through two tools: the *Patto di Collaborazione*[5] (Collaboration Pact) between *Pro.Muovo Association* (citizens), ACER (owner of the property) and the Municipality of Bologna (guarantor); and a contract (comodato d'uso modale[6]) with ACER that the building was only on loan to us: since we couldn't take control of the building free of charge, we agreed to take charge of the renovation work while paying rent, even though it was still quite high considering the dilapidated condition of the building.

GV: *The Municipality, which acts as guarantor, can't intervene on this decision?*

AC: No. The Municipality was fundamental because without its intervention they would never have granted us management of the property, but it was limited to that because all the rest of the rules were dictated by ACER.

In May 2016 we took over the property with the commitment to complete works to the value of €12,000 each year. Since we are architects, we were able to calculate the costs of carrying out the minimum necessary works, managing to achieve €25,000 worth of work during the first year. At that point we could take advantage of a healthy and welcoming place and of the rear garden, the most visible part, to host the first outdoor activities.

We were busy with the self-construction site 2 Sundays per month for a whole year, in an atmosphere of happiness and sharing of knowledge that led us to the creation of a group, to make connections between people who had always lived next to each other but had never met before. The process was also attended by external figures with whom we remained in contact, in particular engineering and architecture students fascinated by this real example of doing their job.

Since the times for the construction site were long, it was important to show that we were recovering that space to turn it into a strong reference point for the district. For this reason, we carried on both the construction and the organization of the first activities in parallel, including a photographic competition on abandoned spaces with a subsequent exhibition inside the building, various concerts, happy hours, etc.

With the inauguration and opening to the public in May 2017, the association of citizens *Pro.Muovo* has become an ARCI (Associazione Ricreativa e Culturale Italiana) club, in order to take advantage of being part of a large circuit in the area of Bologna. This new phase opened two new themes: how to use the new space and what rules to give us for its management.

---

[4] *ACER (Azienda Case Emilia-Romagna)* is a public commercial institution that manages the real estate and provides technical services.

[5] Patto di Collaborazione is a legislative tool that the Municipality and the citizens use together to agree the interventions to take care of, restore and manage common goods.

[6] Comodato d'uso modale basically consists of paying the equivalent of the rent of the building in restructuring works.

In order to establish these new requirements, we started "INavanti", a participatory workshop that provides one afternoon meeting every 3 months, during which we define short and long-term deadlines and objectives.

Meanwhile the project spontaneously continued to succeed and to establish relationships with new people. An opportunity has been the relocation of the km0 market of the *Savena* district to the square in front of the INstabile building: since then, every Friday afternoon, the square hosts a bustle of people, and some of them became an active and integral part of the group. We also organize Wednesday assemblies (two Wednesdays per month), open to everyone, when we outline the daily life of the project trying to be very transparent to the citizens.

It's a very Darwinian process in the sense that not everyone can endure the impact: we are basically a group of volunteers, so everyone invests the time they have, and not everyone does come back but whoever does is committed.

In addition to the management of daily activities, we participated in co-planning laboratories proposed by the Municipality of Bologna for the Participatory Budget (Bilancio Partecipativo[7]).

GV: *How does the Participatory Budget work?*

AC: Citizens can propose projects regarding the territory, that in the beginning are technically selected by the Municipality on the basis of feasibility taking into account the budget and legislation, and then voted on by citizens. Only one project per distict can win (therefore, in the case of Bologna, 6 projects win in total).

Our proposal was to make the square in front of INstabile partly pedestrian, preserving the current mobility and reorganizing the parking lots. We were opposed by a competing project in the same neighborhood, which gave us bad publicity by giving false information about our project. In this way they created a distorted and negative vision about us to people who didn't know us, and moreover their project received the funds.

This is to say that it is a very difficult process in general. We lost the Participatory Budget call but we continued to participate in the neighborhood activities.

The last phase sees the project in a moment of crisis. Thanks to the interest we had in the building, this will be financed in its total restoration through the European funds of the PON[8] (National Operational Plan for metropolitan cities), with the Municipality of Bologna deciding on which properties to invest in. When we took over the building, we knew that for the period of the works, which would have affected the entire building, we would have been obligated to leave, but the timing we indicated was accelerated so we'll have to leave it within a few months. Our request was to have a bridge-building available during the time of the construction site, and we are still negotiating it with the Municipality and with ACER, that for the moment cannot give us guarantees. We are therefore in a quite demanding transition

---

[7] Bilancio Partecipativo is a form of direct participation of citizens in the political life of their city, consisting in assigning a budget for the local authority directly managed by the citizens, who are thus enabled to interact and discuss the choices of the administrations to modify them for their own benefit.

[8] *PON (Programma Operativo Nazionale):* the National Operative Plan for metropolitan cities is a document approved by the European Commission that enables the use of European capital.

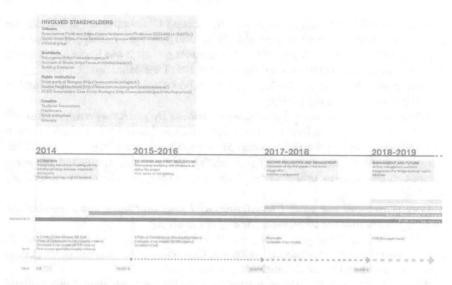

Fig. 3.26 Involved stakeholders and funding, InStabile, Bologna

phase, of re-arrangement of the project, that we obviously don't want to stop during this period (Fig. 3.26).

GV: *At this moment I would say that this project of urban reactivation can lead to 2 different scenarios: Renovation and new use of the building. How do you feel about the experience?*

AC: Certainly, for us it is a successful project, and we have noticed it both on the small scale (for example when we were able to have water and electricity inside the building) and on the big developments that we have been seeing from one year to another, like the increase in citizen participation and in the number of events we hosted. It's a successful case because it obtains positive results, even if obviously it remains in a state of uncertainty and without long-term guarantees. In addition, there is a network of small associations (e.g., *Leila Association* - library of objects) and citizens who have something to teach. There are macro-actors and also a lot of micro-actors.

GV: *In summary, what were the public fundings you received during these 3 years?*

AC: 1. *Coop Adriatica* (€10,000), 2. Municipality of Bologna (€3,000 for a concert and another €3,000 with the second Collaboration Pact, for consumer goods), 3. *Incredibol!* call (€4,000 for a workshop of garden platforms), 4. Neighborhood micro-calls. All the loans we received (except for *Coop Adriatica*) were repayment funds, which means a system based on an initial investment that was later given back to us, through a rather convoluted and complicated process.

The PON funding is managed by the Municipality for the renovation of the building, so we didn't receive anything, not even for the support activities we offer.

A significant breakthrough came with the second Collaboration Pact: while the one we signed the first year was unpaid, the second year the Municipality gave us a reimbursement budget for the purchase of consumer goods. It was important because it marked a change toward us from the perspective of the Municipality, which gave us more confidence because we have proved to be reliable and autonomous, and to have a good impact on the territory.

GV: *Have you ever done self-financing activities? Do you have private sponsors?*

AC: We have done various self-financing events and over time we have become much better at finding the right format to manage them and make them profitable (mostly for events accompanied by aperitifs). We are also sponsored by some companies that give us useful material for free to carry out the renovation of the building.

GV: *Which are the tools ('legislative' and not) that have accompanied the various phases and the implementation of the project?*

AC: 1. Call for funding 2. PON (even if it does not concern us very closely), 3. Participatory budget 4. Collaboration Pact 5. Comodato d'uso modale 6. Workshop Spaces with Urban Center (Municipality of Bologna).

The most important legislative instrument has been the 'comodato d'uso modale' with ACER, which consists of paying the equivalent of the rent of the building in restructuring works. We had to adapt to this tool because ACER (and public administrations in general) is based on 'hardware' rules based on receiving back a quantifiable asset, without being able to give a concrete value to the social impact that is at the base of the project.

The Collaboration Pact is a tool conceived and tested in the Municipality of Bologna which has subsequently spread to other Italian municipalities. Among the options of the Collaboration Pact there is the possibility of managing spaces; not all the collaboration agreements allow it because at their basis there is a strict regulation on the use of common goods. It is basically an agreement between public administration and citizenship (in the case of the management of the spaces it must be a legal group, but for other types of activities it can also be an individual citizen) on the ways of managing spaces according to the *Regulation on the use of common goods.*[9]

This regulation is being developed by the *Spazio Lab* (Urban Center of Bologna), a multidisciplinary office of the Municipality that meets with associations and citizens to define the rules for the management of these transformation spaces through the design.

GV: *In your case, what did the co-design consist of?*

AC: It was a meeting between us, the Municipality and ACER in which we defined the objectives of the project, the tools that the administration could make available, above all in terms of bureaucratic facilities and reference figures to turn to.

---

[9] 'Regolamento dei beni comuni': the Regulation on the use of common goods represents a collaboration tool between the citizens and the municipality to manage shared care of urban common goods. It recognizes in its statute the possibility for citizens to compete directly with the administration of the city, aimed at implementing the constitutional principle of horizontal subsidiarity. Bologna is the model for all the other Italian municipalities that decide to adopt this tool.

GV: *If you had to say what the project's impact on the territory is in a few words, what would you say?*

AC: The impact is on different scales: first of all, the reproducibility of the process. Giving confidence to citizens on the public–private relationship and on the ability of people to take care of spaces is obviously not a visible impact, but those who took part in InStabile try to bring the same model back to other cities.

A civic impact on the level of citizenship: the commitment and passion with which people have worked on the construction site and the development of a project that in reality does not bring any direct income has been repaid both by its success and by the awareness of being able to communicate incisively with the public administration.

Last but not least, the fact of suddenly seeing a building that had been closed for 30 years open and running again has created a strong impact on the social tissue, an empowering of the community that develops a renewed sense of belonging.

GV: *What did you learn about the findings of the process? What are the things that didn't work?*

AC: *Communication findings.* A major criticism from the Municipality side was the lack of a real department to make it easy to understand all the aspects, especially technical, related to the management of the process. So, to be able to carry out such a project it is necessary to be competent in different fields: while in the first phase there was the necessity of specialized figures like architects to cope with technical meetings, in the second phase we needed people who brought their know-how on the constructive, artistic and organizational issues.

System findings. Tools such as the Collaboration Pact cannot in any way affect the 'hardware' system of public administration, with the consequence that for many procedures it is necessary to pay a price that is too high to support what will be the actual use of the building. It becomes really discouraging.

Human Findings. Like every project, the management of human resources is not simple, it's necessary to have a predisposition and empathy, to be a mediator, to be able to often say "no" and to jump into crazy companies. The relational level is very challenging, but it becomes an integral part of your life, which is on the one hand the beautiful part, but on the other it is also the consequence of the inefficiency of the process. If the bureaucracy was leaner and simpler it would undoubtedly be a positive factor, but it would not give way to the community to weave such strong relationships.

Personal findings. INstabile is working because many people (myself included) believe in it, but we are people with our own lives, to which all of our choices are linked. For this reason, being a process that requires a lot of time, dedication and energy at different levels, not all those who are part of it (understandably) want to stay in the long term, with the sad consequence that the project, despite reorganizing, is very much affected in terms of continuity.

GV: *If tomorrow you could start a new INstabile from the beginning, what would you suggest? How would you solve the critical issues you have listed?*

AC: Within public administration there should be the presence of a single contact person who knows all the aspects of the project, to whom the private side can turn without going crazy. There are generally no grants to really give the capacity to these projects to grow (sometimes neither to start). There is no legislative framework, although there will never be a law that regulates this type of use. For example, current laws should include a certain type of temporality in the use of space, or a regulation of 'compromises': keeping to the rules of common sense and safeguarding of people, there are many formal rules that could be easily bypassed but they are not because the law does not provide for the exception to them. This rigidity has a devastating impact on these kinds of projects.

It is very difficult to translate this concept because it's not only a question of simplifying but of 'closing an eye' in certain cases, having simple common sense that unfortunately is not regulated.

3.4.3   Digital Tools, Applications

This part took into consideration the results of the analysis '*ATLAS of emerging practices: being an architect in the 21st Century*'—which involved a selection of 96 practices from 22 European Countries in an online survey divided into 4 sections: organizations, business, media, project.

This section presents the results of the chapter 'media'—which investigates the potential of digital media to improve the work of the architect. On the one hand, the online survey aims to map a list of tools used by the various participants who took part. Also, it is interesting to understand their potential applications through the analysis of various case studies. This questionnaire proposed three questions: first, for the practices to indicate in order of importance the most relevant media they rely on; second, a multiple-choice question regarding the way they are regularly used (e.g., building an online campaign, collaborating/networking, research, etc.); third, in connection to the previous question, asking the practices to give three examples to support their choice. The answers obtained through the first two questions collect quantitative data, highlighting the participants' different purposes and ways of using these tools. The third question collects concrete case studies, which will be explored in this section through a series of interviews with five different participants. The results of the questionnaire were then referred to specific operational needs of the professional activity;—communication and collaboration;—Going public, finding clients, developing projects;—Join forces, collaborate.

Communication and collaboration: the first set of answers collects a long list of digital tools that are used in different ways. Apps, platforms, and web pages are divided into two broad categories: the first one groups together those digital tools that foster communication and facilitate dissemination and research, which often brings external collaboration and clients. The second category relates to the organization and management of the practice's daily activities and communications within the core team or with clients.

Going public, finding clients, developing projects: the first category refers to digital channels such as Facebook or Instagram (just to mention the most common ones), through which everyone can easily communicate and share their work. They are free and easy to use, and evidently so, as seen from most of the participants engaging in these social media channels—influencing the way they communicate the architectural project and the activities developed by the office.

deltastudio (Ronciglione, IT) refers to Facebook and Instagram as tools that allow the dissemination of projects, "Facebook is used to spread the word about the latest projects and news; Instagram is used to build up narratives and to show the making of projects". Being 'connected' and investing time in the dissemination of activities may also generate new commissions. Studio Animal (Barcelona/Madrid, ES) highlights the use of Instagram, increasing the chances of meeting potential clients: "We promote our work via Instagram; some clients have contacted us because of these publications". This same concept has also been expressed by NEAR architecture (Athens, GR/Rome, IT), "We use Facebook to disseminate our work, to present our activities and to open our work to discuss with third parties".

Besides solely for the publicity or the expectation of getting new clients and collaborations, there were other interesting ways some practices used these tools. For example, Space&Matter (Amsterdam, NL) introduces the potential of engaging with their community of followers through their online media channels: "We proactively engage end-users in our design process, by building communities before building (designing) buildings. Through online media, we interact with people and collectives on a frequent basis. We also share our work and knowledge through online media which helps to find partners who are capable of dealing with the same level of complexities".

In addition, in this category, these kinds of tools are often used to promote theoretical investigations and the production of new content such as the search for architectural references, types of projects, or architectural competitions and tenders, as reported by most participants. ENORME Studio (Madrid, ES), for example, "We use Tumblr for personal research projects like Bizarre Columns"; as well as SOUTH architecture (Athens, GR): "Collaboration through Facebook initiated Archpoints, an open house event every year, dedicated to emerging practices, based in Athen's downtown".

Join forces, collaborate: the second set of uses for social media channels is as tools that facilitate the organization of the activities developed within the office such as the division of tasks or the possibility to share and work together on the same files. Among these, the tools most frequently used are Google Drive and Dropbox, referred to 34 and 19 times respectively by the participants of the online survey. Within this second category there are also tools such as Skype (44) that simplify internal communications or long-distance collaboration with other partners, employees, or clients. Google Drive is a widely used tool to organize work files and facilitate collaboration in projects and various activities thanks to the advantage of allowing multiple people to work simultaneously on the same file. These aspects have been highlighted by several practices such as MUTT (London/Liverpool, UK), "*MUTT has roots in both Liverpool and London, with collaborators sometimes based even further afield. Google Drive allows for access to our work from anywhere at anytime*"; and a similar case is made by OFFICE U67 (Aarhus, DK), "*The two directors work* *in two different cities (Oslo and Copenhagen) [...] and only* via *Dropbox can we follow the daily workflow in the office (Aarhus)"*.

Skype and other tools such as Google Hangout and FaceTime were mentioned several times to organize meetings with clients in different locations. krft (Amsterdam, NL), for example, explain that *"Skype allows us to work outside of our direct context, very important. Visual contact is important, above digital messaging. We do many online meetings"*. Taller de Casquería (Madrid, ES) also mention the possibility of using these tools to present their work in public, *"We often lecture and work using Skype, FaceTime, WhatsApp, or similar tools"*.

In the case of both the first and the second category, the usage of some tools is more common than others. The main ones in the first category are Facebook and Instagram; and Google Drive and Dropbox in the second. Together, they offer the benefits of both communication and research possibilities as well as the ease of collaboration and organization. Other responses include using tools such as Pinterest (21), Twitter (18), Vimeo (16), and LinkedIn (8). However, this list is very broad and demonstrates the huge diversity and wide availability of tools intended for more specific purposes. These include Telegram, Trello, Moxtra, HelloAsso, AngelList, Asana, and many others, and although they were mentioned by a minority of participants, they highlight the possibility of developing their own activities quite diversely.

Although the results of the questionnaire show that a only handful of tools are used by the majority of participants, it is interesting to analyze their uses and their possible applications in detail, which are highlighted by a few participants. Many of them pointed out the potential of communication channels such as Facebook and Instagram and open-access tools like Google Drive and Skype, which often translates into opportunities to experiment.

In order to highlight the possibilities offered by these digital tools, this focus chapter presents five interviews with five selected participants. The first interview, with deltastudio (Ronciglione, IT), offers a general overview of the various uses these tools can have in an architectural firm and the benefits they offer, such as new clients. This is followed by interviews conducted with Parasite 2.0 (Milan, IT), ndvr (Antwerp BE), ABACO (Paris, FR), and POOL IS COOL/Collective Disaster (Brussels, BE), exploring the potential of media when doing research, designing, innovating, raising awareness on a specific issue, gathering a larger audience, building communication campaigns, creating new narratives, and more.

Parasite 2.0 (Milan, IT) talks about the project "MAXXI Temporary School: The museum is a school", which experimented with the combination of different technologies and introduces the concept of 'Instagram Architecture', through an intervention in the spaces in front of the MAXXI museum in Rome. ndvr (Antwerp, BE) explain how they used Facebook to launch *"We kopen samen den Oudaan"* ("Let's buy the Oudaan together")—an online campaign for the purchase of an abandoned building, transforming the project into a digital platform to tackle the theme of urban reactivation—; ABACO (Paris, FR) introduces *"La ville qui parle"*—a photographic research project through Instagram that investigates signs and words left on the surface of buildings as new forms of communicating a message; and finally, the project "POOL IS COOL"—created in collaboration with Collective Disaster

(Brussels, BE)—launches a campaign to promote the construction of the first public swimming pool in Brussels by raising awareness among citizens and implementing various types of activities, all done through solid communication.

Interview with Valerio Galeone, Dario Pompei and Saverio Massaro—founders of deltastudio, Ronciglione (IT). Interview conducted by Gianpiero Venturini.

Instagram, Facebook, WhatsApp, Messenger, Skype, Drive, are just a few on an infinite list of digital tools that have revolutionized the way we organize our work, communicate, and design. In this interview, the three founding partners of deltastudio explain the pros and cons of working this way.

Gianpiero Venturini (GP): *In the online survey I invited you to participate in, I asked for each practice to indicate up to three fields that best represented their own firm's interests. In response, you chose the category "Media & Communication". Why is that so?*

Saverio Massaro (SV): First of all, I would make a distinction between the media channels to promote the activity of the studio, like Instagram or Facebook, and the other tools that we use to organize our work or to collaborate with others, such as Drive, Skype, Dropbox, and so on.

Valerio Galeone (VG): Regarding the use of social media channels, we would mention Instagram, Facebook, and our website. Websites are now more and more similar to Instagram, which has become sort of a portfolio of projects, along with other functions. Facebook is used instead for the communication of certain activities, and to stay informed about events and activities of different kinds.

Dario Pompei (DP): Instagram is probably the channel we invest most in, and the one that we use both for research and communication. When we look for images of materials, products and so on, Instagram has replaced Google. For instance, if you are looking for an architecture reference, it becomes much easier to insert a hashtag, and the results can be very accurate.

VG: When we want to say something about our daily activities, we do so by sharing images or videos through the 'stories' section of Instagram. Our Instagram profile collates only images of the finished project. Compared to Facebook, which we use less and less, we receive higher levels of feedback on Instagram.

GV: *Do you use each communication channel differently depending on the content you want to share?*

SM: Exactly, it depends on the content. Facebook is more useful if you want to share some text, whereas Instagram is better for images. These two elements—images and words—define very easily the way we use one or other social media channel.

I think we should also mention tools like WhatsApp or Facebook Messenger. They are replacing emails. There is not much hierarchy among the many types of apps that allow you to send direct messages. Every app is almost as good as the other, and in general, this creates confusion. We talk without distinction with everyone, from family to clients and friends.

DP: An important element linked to the use of these channels lies in the relationship with the client, who is always able to reach you through WhatsApp, Instagram or Facebook Messenger. These channels, for better or for worse, reduce the distance between the architect and the client. Although sometimes the client may invade your personal space if they feel entitled to contact you at any time.

GV: *And then there are all the tools for the internal workflow. What are the advantages of using tools like Drive, Skype, Dropbox, and so on?*

DP: Drive is useful in various ways. Although it has a limited capacity, it's a tool that makes communication much more agile, without the need to attach documents to emails. We try to organize all the folders with drawings and images in a way that we can share all the material with magazines or clients via a simple link. We also use it as our database to organize our work, as well as for the management of tenders and competitions. It has become an indispensable tool to share, comment, and work simultaneously.

SM: In this sense, I believe that the way of working has changed a lot compared to a few years ago. These tools are now accessible to everyone and have simplified many aspects of the studio, including the economic one. Being tools that, in the basic version, are always at no cost, they have reduced the need for that initial investment, which up until a few years ago was mandatory for those who wanted to start working on their own.

GV: *Could you share the pros and cons of the use of various communication tools?*

VG: The time factor is very important. If, on the one hand, some tools reduce the time to be dedicated to certain aspects, the communication requires a high degree of dedication. Planning, managing, editing, and publishing this information through the various channels requires a lot of time and post-production work. Let's think about the case of competitions—once the project is finished, we need to dedicate time to making it readable. In the case of Instagram, for example, you have to prepare the images in the right format, edit the texts, add the captions, and so on. All these operations require time, and therefore investment.

GV: *Has the use of social channels brought you direct benefits?*

DP: Yes, the project 'Martina'—a house we renovated in Caprarola, near Viterbo, in Italy—is very representative in this sense. The owners saw one of our projects on Instagram and they decided to contact us in for the renovation of their apartment. We can say that this commission was brought by an old project post. Once the house was completed, we asked Simone Bossi to take the photographs, and we received a lot of publicity. It worked very well and this demonstrates the importance of investing in the communication of your activity.

Interview with co-founder of Parasite 2.0, Milan (IT)—Eugenio Cosentino. Interview conducted by Gianpiero Venturini.

The project 'MAXXI Temporary School: The museum is a school. A school is a battleground'—winner of the YAP MAXXI competition 2016 in Rome—expresses the potential of the media, both from the point of view of communication and that of digital technology.

Gianpiero Venturini: *Describe the project in a few words.*

Eugenio Cosentino: In 2016 we were selected by the MAXXI museum in Rome to implement the project "MAXXI Temporary School: The museum is a school. A school is a battleground". The museum proposed to work on the theme of sustainability, in the broad sense of the term—economic sustainability of the project, of the materials, the sustainability of the intervention, and of the role of the architect. Besides wanting to talk about all of these themes, we added other questions, such as the desire to work on the museum's public spaces, bringing the contents to the public square and transforming them into a school. The intervention, which aimed to discuss the relationship between people and the environment—the sustainable use of resources and their environmental impact—, took shape through the construction of three different pavilions.

GV: *These are very complex subjects, though connected to each other.*

EC: Let me quote Robert Venturi to answer this question: "We are not for clarity only in terms of simplicity". We want to deepen certain themes in the way that we think is more correct. If simplicity needs to be translated into complexity, fine. The project, therefore, translates into three layers: the physical installation, the contents, and the digital layer.

GV: *How did you bring the digital layer into the installation?*

EC: From the beginning, we wanted to use architecture as a means to communicate the complexity of the themes. From the presentation of the competition to the public event programme, the smartphone app, the use of Instagram, as well as the different ways we have described the project over its various phases, these have all become an opportunity to experiment with new ways of communication and public involvement.

We brought nature into the installation in a digital sense, applying the movie technique called 'green screen' to the surfaces of the pavilions. Together with Silvio Lorusso, Danilo Di Cuia, and Alessio D'Ellena, we created an app where each participant could take selfies, replacing the green color of the surfaces with images of natural places, such as the North Pole, the Sahara Desert, or the Amazonian Forest. Most surfaces were made of recycled car tires—turning one of the most polluting materials on planet Earth into a series of digital green surfaces.

GV: *I'm very interested in the combination of the digital layer and the physical one. Where did you get the inspiration from?*

EC: The concept of 'Instagram architecture' was inspired by the Serpentine Pavilion created in 2015 in London by SelgasCano, which was the most 'instagrammed' architecture of the year. Assuming that architecture is not simply a background for a selfie—but it is the setting within which we develop our lives—we decided to use the so-called 'Instagram architecture' to convey a concrete message.

GV: *Do you have any particular anecdote that represents the media aspects of the project?*

EC: Having proposed such a media project has exposed us to a series of criticisms and sometimes very negative comments. But this was also a central aspect of the experience: through this project we have taken a clear position on issues that bring different opinions together. Treating themes with strong political ideas, one is subjected to the most diverse reactions, which can sometimes result in very negative feedback, but that at the same time have demonstrated the potential of the project to stimulate people by stimulating many reactions.

Interview with Tim Devos, Co-founder of ndvr, Antwerp (BE)

Tim Devos introduces the 'Oudaan' project, which summarizes the potential of media channels as a tool that can be used both to raise awareness about certain urban challenges and to help facilitate participation. These include online campaigns and digital platforms used in order to empower different stakeholders.

GV: *Could you tell me more about the 'Oudaan' building?*

TD: The 'Oudaan' was built by Renaat Braem, one of the most famous modernist architects in Flanders. It was designed as a place to house the city services of Antwerp, but in the end the building was given to the police department. In 2018, the municipality put the building on the market. Besides being a highly questionable operation, there was also no quality control over the process. It was just speculation.

So, we gathered with some friends, saying maybe we should launch a campaign to buy this building together! It was just a crazy idea, but one Friday night after a couple of beers, we created a Facebook page called '*We kopen samen den Oudaan*' (which translates as 'Let's buy the Oudaan together'). It then got a bit out of control and we got about 2,000 Facebook likes over the weekend and a great amount of attention. By Monday, we were getting called by project developers and real estate developers and they were saying, "You guys seem to have a lot of people on board. We have money, let's do it!" It was totally unexpected, but we decided to start making a plan and take the issue very seriously.

GV: *So, just by creating a Facebook page, you raised a lot of attention regarding the Oudaan building. Then what happened?*

TD: In a few days, we were asked by the Flemish Architecture Institute to organize a series of tours around the 'Oudaan'. Then, we decided to use the space to organize a small exhibition and write a manifesto for the building, where it mostly said that people from Antwerp were allowed to invest together in a piece of heritage for the city. A lot of people were very interested in ways of reusing this building. Apart from our team, we asked for a digital development agency to prepare a website to translate our manifesto into a simple, interactive format. You could, for instance, see with very simple animations how the building might change by transforming it into a public infrastructure. That was very important in the initial phase of the project.

GV: *Also, buying a building like this must be very difficult for a small group of people, right?*

TD: Yes, but we started to develop a financial model and we put up the spaces for rent virtually. We had basically four types of spaces where people could subscribe and apply for certain positions and in this way, we would be able to fill up more than 50% of the building. Eventually, we found a real investor who wanted to invest up to 13.5 million. We started to convince the city to stop the selling procedure, but they didn't. In the end, it was a political decision, so we couldn't buy it.

GV: *Are there any lessons you have learnt from this process?*

TD: We decided to continue developing the idea further. What we did in a very spontaneous way could be further developed and structured into a business model. Thus, when we started to develop an online platform, it gave people the necessary structure to group themselves around a particular building and also the possibility of getting in touch with investors. We actually made an initial mock-up version, and we are now developing a fully functioning platform.

GV: *How does the platform work?*

TD: The main concept is the fact that the platform builds communities and expertise, brings people together and gives them access to resources. The idea is that initially, you have a sort of crowd-sourcing map, where everybody might be able to put buildings in it. The next step consists in community-building—citizens can organize themselves around a building of their choice and start dreaming of their possibilities. Then, when a community is big enough and has a clear idea of what they want, they can pitch the idea to the platform, which would support them in different aspects such as organizational structures, financial models, technical issues, and the possibility of exchanging ideas by connecting them with other communities. At this point, our role is very important, in order to connect with the experts, coaching, building up business models, and so on. The last stage would be the funding and then the platform would connect investors and communities.

GV: *It seems that media is a very important tool to build communities and also to bring attention to urgent urban challenges. Is that so?*

TD: With our platform, we wanted to make a statement about the need for alternative forms of development. A part of the platform is our responsibility to raise awareness of spatial challenges and indeed, the media is very important in helping to bring attention to these issues and helping make these issues more comprehensible for everyone.

Interview with Alice Braggion and Alessandro Carabini-founders of ABACO, Paris (FR). Interview conducted by Gianpiero Venturini.

ABACO investigates the potential of Instagram in the field of design and communication. Mapping, researching, and the dissemination of content become important key points to understand both the project 'La ville qui parle' and the studio's approach.

Gianpiero Venturini: *'La ville qui parle' collects interesting insights into the relationship between digital media, research, and communication. Could you tell me more about the project?*

Alice Braggion: *'La ville qui parle'* is a photographic research project we launched in 2017 at the fourth edition of the Festival New Generations in Rome, and which then took shape in the Unfolding Pavilion at the 16th International exhibition of Architecture Biennale in Venice. It is a project that stems from our interest in communication and digital media, and at the same time translates into a project about our obsession with everything that is related to words.

Alessandro Carabini: Digital media becomes a useful tool for investigating the relationship between the built environment and its inhabitants. The project looks for signs, words, and texts left on the surfaces of buildings—signs and words that often change very quickly. The speed at which the urban environment and the city evolves day-to-day can be found in the speed of posting an image through social media, like Instagram—one of the main tools that we use to promote the project.

Also, the relationship between the city and the user is very important. We use the expression 'city-empathy' to express the particular form of interaction—an emotional one—between the city and the person when they decide to leave a mark, communicate a message, or provide a reaction.

GV: *How is the project structured and how does it work?*

AB: The use of social networks and the participatory aspect are both very important. The most suitable digital channel to translate the various components of the project is Instagram, both for the speed in posting images and because it is based on communication through both images and texts. The open dimension translates into the possibility for everyone to participate by sending photos that reflect the objectives of the project. Now, we ask people from different countries who are interested in contributing, to send us an image accompanied by a short text—a sentence that represents the subject.

AC: We ask for them to precisely indicate the coordinates and the moment when the photograph was taken. It often happens that the same place photographed at two different moments presents different conditions. Time and space are therefore two fundamental components.

GV: *This aspect, the correspondence between space and time, is linked to the relationship between the physical and digital world, where the project develops. How do you explain this relationship?*

AC: When we were developing a spin-off of the project in Venice, we immediately realized that one of the most interesting aspects was the need to deal with very different contexts. Before going there in person, we decided to use Google Street View to take a look at some parts of the city, remotely. Google Street View images were taken a couple of years before, and when we went there first-hand, we realized that many of those signs had changed completely and had given birth to new layers, where multiple dialogues took place.

GV: *Returning to the theme of communication in the broad sense of the word, what importance do you give this in your research?*

AC: One important project for us is the 'Palomar' interactive installation—a communication device—where we extensively used images and light to connect with people. Thus, it acquires a wider dimension thanks to the use of digital tools that propose a different narrative by adding new levels of meaning. Constructed out of digitally fabricated modules, the 'Palomar' contains a message on its surface that becomes communication itself. Architecture is communication—not only in the way we live our daily practice but also and more importantly, in its strong influence on our work through digital media.

Interview with Louisa Vermoere and Paul Steinbrück—co-founders of 'POOL IS COOL'.

POOL IS COOL aims to raise citizens' awareness around the theme of water, bringing the city's first public swimming pool to Brussels through a series of online and offline actions.

GV: *How did the 'POOL IS COOL' project start?*

LV: I arrived in Brussels in 2011 and it was a super-hot summer. I was like, "where do we swim in summer?" "There must be a place!" So, naïve as I was, I started looking on Facebook and there was no reply, zero. It was funny because Paul, who was not even a friend yet, was actually posting the same thing on Facebook. And there, we found each other.

PS: After a couple of meetings, by then joined by other enthusiasts, the 'POOL IS COOL' name popped up just as a placeholder for us to talk about it. And in Brussels—a city of a multitude of languages—catchphrases like 'POOL IS COOL' are pleasant and easier as a name that needs to be translated into all the different official languages.

GV: *Why did you start 'POOL IS COOL'?*

LV: We, citizens in Brussels, wanted to have a place to swim in the open air. And now, this doesn't exist.

PS: Yes, we think that it has a real, important benefit for the quality of life of all the citizens of Brussels. And quality of life is very broad, for some people it's just pleasure; for some others, it's sport; for some, it's the social component like meeting other people.

GV: *How did the idea evolve over the years?*

PS: During the first year, in 2015, we basically went swimming. It was a sort of low-key public disobedience, we jumped into fountains and ponds with friends, just to see what it was like. And we took pictures of these events.

LV: And those pictures were very important.

PS: This material we gathered could immediately be used for communication once 'POOL IS COOL' went public with these activities and it would not remain just a vague, theoretical idea. That was at the beginning of 2016, also the year we became an official organization.

GV: *Would you define this project as a campaign?*

PS: 'POOL IS COOL' is indeed a campaign, in order to convince politicians, administration, and any related organizations to influence the public and steer political debates towards what we think is a favorable direction. The campaign builds both on an offline presence and an online one, which is equally important. We have a website, we are on all the social media platforms, and we have petitions running.

PS: While the activities and events are very important, online communication is also important to frame some things that were not as successful in real life. For example, there's a very funny story to exemplify this. There was one event where we transformed a fountain in the centre of the city into a beach swimming pool. It was a fantastic summer, but it just happened to be raining on that day. So, it was a bit sad for us, but we took a lot of photos anyway and posted them on our social media channels. It was a fantastic event, but it just needed a bit of reframing in the online communication to the public to avoid the inevitable question in Brussels, "Isn't the weather too bad for outdoor swimming anyway?".

GV: *It is always very difficult to reach people and get them to participate in activities. How do you do that?*

LV: This combination of digital and physical is crucial. For instance, in 2015, we made an installation in a place that wasn't the most central. It made so many different groups of people come together around a pool—which was not really a pool, more like a water mirror with a bit of a deeper part—where kids really had a lot of fun. All the groups liked the element of surprise.

PS: One year after, we were asked by BOZAR, the Centre for Fine Arts, in the center of Brussels, if we could build an installation on their doorstep. They realized that in summer they had a problem getting people into their museum. So, they thought that they'd combine it with something on the outside—a space to hang out, to relax. Since it is a museum or an art place with a mission to raise awareness to steer public debates with their program, they couldn't just put a pool for fun. They needed a mission behind it, something like a social ambition and that's why they asked us to be involved, not just to build a pool, but to make this pool a message for our mission. That's how this little pool—literally just a container—became the biggest public outdoor pool in Brussels.

LV: And after the Biggest Pool and other noteworthy activities and events, our social networks gathered double the following they had before. There were so many people that all of a sudden started contacting us. There were so many likes on Facebook, there were more people subscribing to our newsletters. Newspapers were contacting us more often for interviews and we were asked to give our opinions on certain topics that politicians were discussing.

GV: *Do you think that you are producing concrete results through your campaign?*

PS: I think so. Thanks to our campaign, this topic in Brussels can no longer be avoided. A lot of politicians are talking about it, of course, nothing promised yet, nothing really being said. But behind the scenes, we feel that there is something really happening.

3.5   Field Experience, Rigenerare Il Ciano

The project *Rigenerare il Ciano*, led by the New Generations Cultural Association, in collaboration with ACER Piacenza—Azienda Casa Emilia Romagna, and the Cultural Association Urban Curator TAT, consisted of an urban reactivation process in the 'San Sepolcro' neighborhood—better known by local inhabitants as 'Il Ciano'. The project was developed through a participation process with the involvement of the local inhabitants, institutions and the Politecnico university, Polo di Piacenza. The project was developed over three phases, which can be summarized as follows:

(a) a phase consisting of research and territorial analysis; (b) a design workshop, accompanied by a program of activities and seminars; (c) the realization of a temporary pavilion through a self-construction workshop and with the involvement of various local stakeholders.

Phase A—Research and territorial analysis

The first phase of the project consisted of conducting territorial analysis and meetings with experts, as well as organizational activities in preparation for the second phase of the project. The aim of this phase was to analyze the characteristics of the project area and to lay the foundations for establishing a relationship with the inhabitants and local institutions.

Between December 2018 and March 2019, the research group carried out four site inspections, through which it was possible to gather information related to the spatial configuration of the neighborhood and to approach the community of residents, represented by more than 80 families of different nationalities. As far as the territorial analysis is concerned, part of the activities focused on the search for bibliographic material and on the creation of a photographic map related to the condition of the project area. During the site visits, some interviews were conducted with a selection of families: The interviews were a useful tool both to better understand some of the problems that plague the neighborhood (lack of maintenance, micro crime, noise, etc.) and to establish an initial dialogue with our direct interlocutors, the residents of the neighborhood. It was a key moment thanks to which it was possible to lay the foundations of a trusting relationship that was consolidated during the project.

In parallel, some meetings were held with the experts involved in the project, with whom the further phases could be planned, also on the basis of what emerged from the analysis of the site. Daniele Fanzini, expert in urban regeneration and participatory practices, was called in to plan the participatory phases, involving the local community and the Politecnico di Milano in the workshop phases; various representatives of the cultural association Urban Curator TAT were involved in the design workshops held at the Politecnico di Milano—Polo di Piacenza, during the second phase of the project (April 2019); Giuseppina Civardi, social mediator of ACER Piacenza, expert of the San Sepolcro district, accompanied the different working groups during the four site visits. The figure of the social mediator has proved to be very important, acting as a filter for the inhabitants of the neighborhood. Finally, Patrizio Losi, president of ACER Piacenza, and Fabrizio Cavanna, director of ACER Piacenza, who, with their institutional role, rationalized some bureaucratic requirements that were necessary for the realization of the subsequent phases of the project. These meetings took place monthly during the first three months of the project and proved to be very important as they facilitated the planning of the next steps of the project. From this experience, the need to create a flexible project plan from the beginning, open to possible changes resulting from different unpredictable situations, becomes clear.

The activities of the project were initially planned through a detailed schedule based on the Gantt model, but this had to be readjusted as the project progressed. In the first week of March 2018, Giuseppina Civardi and Gianpiero Venturini carried out a further inspection, during which it was possible to talk to some tenants from the neighborhood, to whom the project plan and the initiatives that would be carried out in the following weeks were made known. From this meeting, where it was possible to present some ideas for the realization of the self-build intervention, emerged the need to work on the central areas of the neighborhood (Fig. 3.27).

Phase B—Design workshop and seminars

During the course of the project and after some meetings with the tenants and other experts involved in the first phase of the initiative, the decision was taken to implement the project proposal by holding a design workshop, which took place at the Politecnico di Milano - Polo di Piacenza. Initially, the project proposal foresaw the implementation of a self-construction workshop, to be realized in the second phase of the project, after meetings with the residents. Given the particular interest of the residents of the neighborhood, it was considered necessary to implement the proposal with an intermediate step: the holding of a design workshop, carried out together with a selection of students from the Faculty of Architecture of the Politecnico di Milano

Fig. 3.27 Tryptic, San Sepolcro Neighborhood, Piacenza

- Polo di Piacenza. The aim of the workshop was to produce a series of targeted proposals to be presented to the residents of Ciano.

A three-day design laboratory was held in the classrooms of the Politecnico di Milano, with the aim of collecting ideas from students and providing additional tools for dialogue with residents. The design brief was to think of a temporary pavilion with a maximum extension of 100 square meters. The idea of the pavilion emerged from the initial meetings with the residents of the neighborhood. It was intended as a multifunctional space that could be adapted to different configurations, based on the different functions proposed by the residents of the San Sepolcro neighborhood. The lab was designed with exercises aimed at direct communication with the residents and to help them understand the dynamics of a self-construction project: cheap and readily available materials, easy-to-execute construction techniques and the temporary nature of the project. The program was developed through a range of activities including: a neighborhood walk conducted with the students, to discuss on site the dynamics and opportunities offered by the shared spaces of the neighborhood. During the afternoon activities, participants were organized into separate groups of 5 to 6 people. An initial exercise was then conducted in the classroom with the aim of determining the position of the pavilion: despite the initial feedback from the residents who had indicated the central courtyard as a suitable place to build the pavilion, the students were invited to design alternative proposals and open the field to any solution. Using diagrams, sketches and a 1:200 scale plan, the students began to shape their pavilion and explore how the structure could dialogue with and extend the existing spaces.

The second and third days were devoted to building a 1:20 scale model. The aim of the exercise was to show the residents the possibilities offered by the project in a vivid and intuitive way. It was therefore decided to build the models on a larger scale than expected and to use unusual techniques and materials in the academic field.

Conferences, seminars and presentations involving experts were also organized in parallel with the workshop activities: Maria Vittoria Capitanucci gave a lecture on the architecture of working-class neighborhoods in Italy; Gianpiero Venturini and Giovanni Castaldo talked about the tools for urban reactivation and presented a selection of international case studies; Matteo Pettinaroli presented the work of the Needle—Urban Acupuncture research project, which addresses the issue of urban reactivation with an interdisciplinary approach; Mario Paris and Alice Buoli, both professors at the Politecnico di Milano, presented different methodologies; Annalaura Ciampi and Luca Vandini, co-founders of the Kiez.Agentur, presented InStabile—Community Creative Hub, a project they are responsible for; the Spanish studio VIC—Vivero de Iniciativas Ciudadanas in turn spoke about MARES Madrid, a European project for the reactivation of 4 buildings on the outskirts of Madrid.

On the last day of the workshop, a presentation of the results was attended by all parties involved in the process: New Generations, ACER, Urban Curator TAT, institutional representatives of the Politecnico di Milano and the Municipality of Piacenza, a selection of residents from the San Sepolcro district and approximately 40 Politecnico students who participated in the realization of the project proposals.

The event was developed in three phases:

- the presentation of the initiative, made by the scientific director of the project, Gianpiero Venturini, with the participation of the President of ACER, Patrizio Losi, and the expert in urban regeneration, Daniele Fanzini;
- a series of six presentations of the ideas developed by the students of the Politecnico di Milano. During each presentation, lasting about 6–8 min, the students presented the main characteristics of each project proposal, the objectives, and the functional and technical characteristics for the construction of the pavilion;
- an open discussion between the various participants and the residents of the neighborhood, highlighting potentials and criticalities of each proposal.

The meeting was useful for the progress of the project, collecting feedback from the various users involved, and planning the next phases. The first important step before the construction of the pavilion was to create a questionnaire, which aimed—in a simple and timely way—to collect the opinion of the neighborhood inhabitants who were unable to participate during the public event. The questionnaire was shared with the approximately 200 families who live in the same number of houses in the San Sepolcro district: the answers were used to draw up a definitive project proposal for the construction of the pavilion (Figs. 3.28, 3.29, 3.30 and 3.31).

Fig. 3.28 Rigenerare il Ciano, meeting with the inhabitants

Fig. 3.29 Rigenerare il Ciano, presentation of the proposals

Fig. 3.30 Rigenerare il Ciano, presentation of the proposals

Fig. 3.31 Rigenerare il Ciano, presentation of the proposals

Phase C—Realization of the intervention

The last phase of the project consisted in the realization of the intervention in the central courtyard of the neighborhood: a temporary, multifunctional pavilion inserted in the context of the San Sepolcro district and donated to the community of inhabitants. Although the pavilion was designed as a temporary structure, to last a few months, the intervention remained inside the courtyard for over a year: neighborhood tenants and associations working in the local area such as Caritas, welcomed the structure, planning extra-project activities through a cultural program that has aroused the interest of the community of citizens.

Starting from the indications received from the inhabitants and from the students who participated in the workshop, the final version of the pavilion took into consideration the many aspects that emerged during the first two phases of the project: from the environmental conditions to the limitations imposed by the willingness to build the pavilion through a shared construction process, which must therefore be built by non-professionals and in a limited period of time. Even the graphics produced in order to carry out the construction process are different from those usually required for an architectural project: it was decided to simplify construction and make the individual construction phases explicit and efficient, so as to be able to easily coordinate the construction work.

The pavilion was therefore conceived as an open and permeable structure, capable of accommodating various functions. Both conceptually and formally, the pavilion replicates the internal configuration of the homes in Il Ciano: different areas of the pavilion have been designed to recall the domestic spaces that characterize the homes of the inhabitants, with an equipped living area, a corridor, two bedrooms and a balcony. This idea emerged during the making of the video documentary, during which it was possible to visit numerous houses and study their configuration.

Despite the great cultural diversity represented by the many families who live in Il Ciano, during the various inspections carried out to make the video recordings, a common feature emerged: the configuration of the domestic space is almost always the same, and each inhabitant decorates it, personalizing the space in different ways. It was therefore thought that the pavilion should recall a family environment, offering a configuration similar to that found in individual homes. The structure was built with standard materials, such as raw spruce strips with square section and marine pine panels, easy to find in any carpentry or hardware store, and built with a few easy-to-use electrical tools.

The pavilion was entirely built thanks to the help of a group of 30–40 people: local and international guests who contributed to the organization of the construction site, students of the Politecnico who had took part in the design workshop, and who offered to follow the construction phases, representatives of the various associations involved, and neighborhood residents who, intrigued, offered to participate in the construction phase.

The students followed the construction phases throughout the workshop: from the processing of the materials (cutting, sanding, painting), to the final assembly and decoration. The pavilion was welcomed by the inhabitants of the neighborhood with enthusiasm: after a first moment of distrust, many of the inhabitants approached and appreciated the new building. At the end of the works, some residents suggested inaugurating the structure, organizing a small party. The pavilion was inaugurated during the month of July 2019, with the participation of many families, children, the elderly, local inhabitants and other visitors, who found themselves using the new space (Figs. 3.32, 3.33 and 3.34).

Fig. 3.32 Rigenerare il Ciano, project realized in collaboration with the inhabitants

Fig. 3.33 Rigenerare il Ciano, project realized in collaboration with the inhabitants

Fig. 3.34 Rigenerare il Ciano, project realized in collaboration with the inhabitants

Reference

Venturini G, Venegoni C (2016) Re-Act: tools for urban re-activation, vol 1. Deleyva Editore, Monza

Printed in the United States
by Baker & Taylor Publisher Services